Electrical Machines

Other Macmillan titles in Electrical and Electronic Engineering

B. R. Bannister and D. G. Whitehead, *Fundamentals of Modern Digital Systems*
G. B. Clayton, *Data Converters*
J. C. Cluley, *Electronic Equipment Reliability, second edition*
J. C. Cluley, *Transducers for Microprocessor Systems*
R. F. W. Coates, *Modern Communication Systems, second edition*
A. R. Daniels, *Introduction to Electrical Machines*
C. W. Davidson, *Transmission Lines for Communication*
M. E. Goodge, *Semiconductor Device Technology*
B. A. Gregory, *An Introduction to Electrical Instrumentation and Measurement Systems, second edition*
L. F. Lind and J. C. C. Nelson, *Analysis and Design of Sequential Digital Systems*
Paul A. Lynn, *An Introduction to the Analysis and Processing of Signals, second edition*
A. G. Martin and F. W. Stephenson, *Linear Microelectronic Systems*
J. E. Parton, S. J. T. Owen and M. S. Raven, *Applied Electromagnetics, second edition*
M. Ramamoorty, *An Introduction to Thyristors and their Applications*
Douglas A. Ross, *Optoelectronic Devices and Optical Imaging Techniques*
Trevor J. Terrell, *Introduction to Digital Filters*
M. J. Usher, *Sensors and Transducers*
G. Williams, *An Introduction to Electrical Circuit Theory*

Electrical Machines

An Introduction to Principles and Characteristics

J.D Edwards

*Lecturer in Electrical Engineering
School of Engineering and Applied Sciences
The University of Sussex*

Second Edition

Macmillan Publishing Company

NEW YORK

© J. D. Edwards 1986
Copyright © 1986 by Macmillan Publishing Company
A Division of Macmillan, Inc.

All rights reserved. No part of this book may be reproduced or transmitted in any form or by any means, electronic or mechanical, including photocopying, recording, or by any information storage and retrieval system, without permission in writing from the Publisher.

Macmillan Publishing Company
866 Third Avenue, New York, NY 10022

Collier Macmillan Canada, Inc.

Printed in Hong Kong

printing number   year
1 2 3 4 5 6 7 8 9 10   6 7 8 9 0 1 2 3 4 5

Library of Congress Cataloging in Publication Data
Edwards, J. D.
Electrical machines: an introduction to
principles and characteristics
1. Electrical machinery
I. Title
621.31′042   1986   86-061289

ISBN 0-02-948060-4

The cover photograph shows a 20 MW, 1200 rev/min, 60 Hz brushless synchronous motor used to drive a high-speed compressor. (Courtesy of GEC Large Machines Ltd, Rugby, England.)

Contents

Preface	*ix*
Notation, Units and Symbols	*xi*

1 General Principles   **1**
- 1.1 Introduction   1
- 1.2 Conductor moving in a magnetic field   2
- 1.3 Electromagnetic induction   9
- 1.4 Electromagnetic forces   14
- 1.5 Magnetic materials   23
- 1.6 The magnetic circuit   28
- 1.7 Permanent magnets   38
- Problems   41
- References   43

2 Direct Current Machines   **45**
- 2.1 Introduction   45
- 2.2 Fundamental principles   46
- 2.3 Energy conversion and losses   58
- 2.4 DC generators   61
- 2.5 DC motors   64
- 2.6 Dynamic characteristics of DC machines   72
- 2.7 Special machines   75
- Problems   77
- References   79

3 Alternating Current Systems   **80**
- 3.1 Introduction   80
- 3.2 Generation of sinusoidal alternating voltages   80
- 3.3 Polyphase systems   82
- 3.4 Transformers   86
- Problems   100
- References   102

4 Introduction to AC Machines — 103
- 4.1 Introduction — 103
- 4.2 Distributed windings and the airgap magnetic field — 104
- 4.3 Combination of sinusoidally distributed fields — 108
- 4.4 Torque from sinusoidally distributed windings — 111
- 4.5 The rotating magnetic field — 114
- 4.6 Voltage induced by a rotating magnetic field — 117
- 4.7 Multi-pole fields — 122
- 4.8 Introduction to practical windings — 124
- Problems — 128
- References — 130

5 Synchronous Machines — 131
- 5.1 Introduction — 131
- 5.2 Phasor diagram and equivalent circuit — 132
- 5.3 Synchronous machine characteristics — 136
- 5.4 Salient-pole synchronous machines — 143
- 5.5 Stepper motors — 146
- Problems — 155
- References — 156

See the difference on p 114

6 Induction Machines — 157
- 6.1 Introduction — 157
- 6.2 Electromagnetic action — 159
- 6.3 Equivalent circuit — 161
- 6.4 Induction machine characteristics — 166
- 6.5 Speed control of induction motors — 173
- 6.6 Single-phase induction motors — 177
- 6.7 Linear induction motors — 180
- Problems — 182
- References — 183

7 Generalised Machine Theory — 184
- 7.1 Introduction — 184
- 7.2 DC machine equations — 186
- 7.3 AC machine equations — 188
- 7.4 General equations — 191
- 7.5 Applications — 193
- 7.6 Conclusion — 200
- Problems — 201
- References — 202

Appendix A: Airgap Field Components and the Maxwell Stress	*203*
Appendix B: Three-phase to Two-phase Transformation	*206*
Appendix C: Dynamic Braking of an Induction Motor	*210*
Bibliography	*215*
Answers to Problems	*217*
Index	*219*

Preface

This new edition retains the original aim of the first edition: to give a short modern account of the principles of electrical machines in a form accessible to the non-specialist, yet with sufficient rigour to provide a basis for more advanced studies. Much of the book has been rewritten to reflect the latest developments in machines and power electronics, and a new treatment of AC machine theory is presented. The chapter on generalised machine theory has been revised, and it now includes an example of the computer solution of a practical machine problem. Quite deliberately, there is no treatment of power electronic circuits. This is now such a large field that it is impossible to do it justice in a book of this size. Power electronic control of machines is discussed in chapters 2, 5 and 6, and there are suggestions for further reading in the Bibliography at the end of the book.

Electrical machines form only part of a large class of electromechanical devices which all depend on a small group of basic physical principles. Two approaches are common in the literature: the abstract unified theory, derived from the underlying physics; and the piecemeal treatment of the different devices, with little regard for common principles. Although suitable for the specialist, both of these approaches are unsatisfactory for the non-specialist. In developing the theory of machines I have chosen a direct physical approach, based on the interaction of currents and magnetic fields. This gives a clear physical picture of the operating principles of each type of machine, and unity is preserved by a common method of analysis.

The treatment is fully quantitative, and the requirements of rigour and simplicity have been met by taking the simplest model of a machine which will demonstrate the essential features of its operation. Departures from the ideal are mentioned only briefly, since these form the subject of specialised study. Likewise the varieties of practical windings are not treated; but the functions of machine windings are fully explained.

AC machines generally give students the most difficulty; a feature of this book is a new approach to AC machines, introduced in chapter 4 and developed in chapters 5 and 6. Traditionally, the concepts of MMF and flux dominate the theory; these are useful concepts, but they make the action of the machine difficult to visualise because they are one step removed from the fundamental current and field. I have therefore based the theory on the concepts of current density and flux density. These concepts give an immediate picture of the electromagnetic action; they also lead directly to the rotating field principle, the torque equation and the equivalent circuit.

In the final chapter there is an introduction to generalised machine theory, which is indispensable for advanced work such as the transient behaviour of AC machines. Since rotating electrical machines are components of most engineering systems, it seems desirable to give at least an introduction to this powerful method of analysis; but the chapter may well be omitted at a first reading.

The reader is assumed to have an elementary knowledge of electromagnetism, circuit theory, vectors, matrices and differential equations. Background material is listed in the Bibliography at the end of the book. Space has limited the number of worked examples and problems for solution. In general the problems at the ends of the chapters are not straightforward, and they extend the material of the chapter. Those who require more problems and worked examples will find them in the books listed in the Bibliography.

I wish to thank the Controller of Her Majesty's Stationery Office and the Longman Group for permission to reproduce figures 1.36 and 2.12 respectively, and several manufacturers (acknowledged in the text) for photographs of machines. I am grateful to reviewers of the first edition, and my university colleagues, for suggestions; to many second-year engineering students for encouraging me to find better ways of presenting machine theory; and to my wife for word-processing the text.

J. D. Edwards

Notation, Units and Symbols

SI units are used throughout the book, and the recommendations of the British Standards Institution [1] are followed for unit names and symbols. With electrical quantities, the usual convention is followed in denoting time-varying values by lower-case symbols and steady values or magnitudes by upper-case symbols. Three kinds of vector quantity occur in the book: space phasors, time phasors and three-dimensional vectors. Bold-face type is used for all these quantities, in line with normal practice, for example I, B. In some cases the same symbol may be used for a field vector or a space phasor; the context will always show which meaning is intended. Bold-face type is also used for matrices.

In general the recommendations of the Institution of Electrical Engineers[2] and the British Standards Institution [3] have been followed for symbols, abbreviations and subscripts. One exception is the symbol for linear current density; the recommended symbol A seems a poor choice because it is the established symbol for magnetic vector potential, and distinct symbols for both quantities are needed in advanced work. The symbol chosen is K, which follows Stratton [4] and some current practice.

There is one area of notation in which no uniformity exists: the choice of symbols for the infinitesimal elements in line, surface and volume integrals. After careful consideration the notation adopted is that of Stratton [4], in which ds, da and dv denote the elements of path length, area and volume respectively. This is a consistent and convenient notation; it justifies the use of the subscript s for the tangential component of a vector; it releases capital letters for designating finite regions or quantities; and there is no confusion in advanced work between the symbols for length or area and the Poynting vector S or the vector potential A.

The following lists refer only to the usage in this book. Further information will be found in the references already cited and in Massey [5].

List of principal symbols

Symbol	Quantity	Unit	Unit symbol
A	area	square metre	m^2
a	radius	metre	m
$B, \mathbf{B}$	magnetic flux density	tesla	T
B	flux density phasor	tesla	T
C	capacitance	farad	F
$da, d\mathbf{a}$	element of area	square metre	m^2
$ds, d\mathbf{s}$	element of path length	metre	m
dv	element of volume	cubic metre	m^3
$E, \mathbf{E}$	electric field strength	volt/metre	V/m
$E; \mathbf{E}$	excitation voltage phasor; RMS magnitude	volt	V
$E; \mathbf{E}$	induced EMF phasor; RMS magnitude	volt	V
E, e	electromotive force	volt	V
$F, \mathbf{F}$	mechanical force	newton	N
F	magnetomotive force	ampere	A
F_x	x component of force	newton	N
$f, \mathbf{f}$	force per unit volume	newton/metre^3	N/m^3
f	frequency	hertz	Hz
g	airgap length	metre	m
$H, \mathbf{H}$	magnetising force	ampere/metre	A/m
$I; \mathbf{I}$	current phasor; RMS magnitude	ampere	A
I_0	no-load current	ampere	A
I_{01}	loss component of I_0	ampere	A

NOTATION, UNITS AND SYMBOLS

I_{0m}	magnetising component of I_0	ampere	A
I, i	current	ampere	A
J, J	current density	ampere/metre^2	A/m^2
J	moment of inertia	kilogram metre^2	kg m^2
j	$\pi/2$ operator, $\sqrt{(-1)}$	—	—
K, K	linear current density	ampere/metre	A/m
K	current density phasor	ampere/metre	A/m
K	DC machine constant	$\begin{cases}\text{volt second/weber radian}\\ \text{newton metre/ampere}^2\\ \text{volt second/ampere radian}\\ \text{newton metre/weber ampere}\end{cases}$	V s/Wb rad N m/A^2 V s/A rad N m/Wb A
K_a	armature constant	weber/ampere	Wb/A
K_f	field constant	newton metre/tesla^2	N m/T^2
k	torque constant	henry	H
L	self-inductance	henry	H
L_m	magnetising inductance	henry	H
l	leakage inductance	metre	m
l	length	henry	H
M	mutual inductance	—	—
m	number of phases	—	—
N	number of turns	—	—
n	Steinmetz index	—	—
n	turns ratio	watt	W
P	power	watt	W
P_e	electrical power	watt	W
P_m	mechanical power	1/second	1/s
p	d/dt operator	—	—
p	number of pole pairs		

Symbol	Quantity	Unit	Unit symbol
p_e	eddy current power loss per unit volume	watt/metre³	W/m³
p_h	hysteresis power loss per unit volume	watt/metre³	W/m³
q	electric charge	coulomb	C
R	resistance	ohm	Ω
R_e	equivalent total resistance	ohm	Ω
R_c	core loss resistance	ohm	Ω
r	radius	metre	m
S	reluctance	ampere/weber	A/Wb
s	fractional slip	—	—
T	torque	newton metre	N m
T_m	mechanical output torque	newton metre	N m
T_θ	torque associated with angle θ	newton metre	N m
t, t	stress	newton/metre²	N/m²
t	time	second	s
U	magnetic potential difference	ampere	A
u, u	linear velocity	metre/second	m/s
$V; V$	voltage phasor; RMS magnitude	volt	V
V, v	terminal voltage, electric potential difference	volt	V
W	energy, work done	joule	J
W_e	electrical energy	joule	J
w_h	hysteresis energy loss per unit volume	joule/metre³	J/m³

NOTATION, UNITS AND SYMBOLS

W_m	magnetic stored energy	joule	J
X	reactance	ohm	Ω
X_e	equivalent total reactance	ohm	Ω
X_m	mutual or magnetising reactance	ohm	Ω
X_s	synchronous reactance	ohm	Ω
x	leakage reactance	ohm	Ω
Z, Z	impedance	ohm	Ω
Z	maximum linear conductor density	1/metre	1/m
α, β, γ	general angles	radian	rad
δ	load angle, angle between magnetic field axes	radian	rad
δ	depth of penetration	metre	m
ϵ	voltage regulation	—	—
η	efficiency	—	—
θ	angular displacement, rotor angle	radian	rad
Λ	permeance	weber/ampere	Wb/A
λ_h	Steinmetz coefficient	—	—
μ	absolute permeability = $\mu_0 \mu_r$	henry/metre	H/m
μ_0	magnetic constant = $4\pi \times 10^{-7}$	henry/metre	H/m
μ_r	relative permeability	—	—
ρ	charge per unit volume	coulomb/metre^3	C/m^3
ρ	resistivity = $1/\sigma$	ohm metre	Ω m
σ	conductivity = $1/\rho$	siemens/metre	S/m
τ	time constant	second	s
τ_{em}	electromechanical time constant	second	s

Symbol	Quantity	Unit	Unit symbol
Φ	magnetic flux phasor	weber	Wb
ϕ	magnetic flux, flux per pole	weber	Wb
ϕ_l	leakage flux	weber	Wb
φ	phase angle, angular displacement	radian	rad
ψ	angular displacement	radian	rad
Ω, ω	angular velocity	radian/second	rad/s
ω	angular frequency = $2\pi f$	radian/second	rad/s
ω_r	rotor angular velocity	radian/second	rad/s
ω_s	synchronous angular velocity	radian/second	rad/s

General subscripts

a	armature
av	average
f	field
m, max	maximum value
n	normal component
r	radial component
s	tangential component
1, 2	primary, secondary; stator, rotor
α, β	two-phase quantities
a, b, c	three-phase quantities
d, q	two-axis rotor quantities
f, g	two-axis stator quantities

Decimal prefixes

10^6	mega	M
10^3	kilo	k
10^{-2}	centi	c
10^{-3}	milli	m
10^{-6}	micro	μ

Abbreviations

AC	alternating current
DC	direct current
EMF	electromotive force
MMF	magnetomotive force
rev/min	revolutions per minute
rev/s	revolutions per second
RMS	root-mean-square

References

1. BS 3763, *The International System of Units (SI)* (London: British Standards Institution, 1976).
2. IEE, *Symbols and Abbreviations for Electrical and Electronic Engineering*, 3rd ed. (London: Institution of Electrical Engineers, 1980).
3. BS 5775, *Specification for Quantities, Units and Symbols* (London: British Standards Institution, 1982).
4. J. A. Stratton, *Electromagnetic Theory* (New York: McGraw-Hill, 1941).
5. B. S. Massey, *Units, Dimensional Analysis and Physical Similarity* (London: Van Nostrand Reinhold, 1971).

1 General Principles

1.1 Introduction

In 1820 Oersted discovered the magnetic effect of an electric current, and the first primitive electric motor was built in the following year. Faraday's discovery of electromagnetic induction in 1831 completed the foundations of electromagnetism, and the principles were vigorously exploited in the rapidly growing field of electrical engineering [1]. By 1890 the main types of rotating electrical machine had been invented, and the next forty years saw the development of many ingenious variations, along with refinement of the basic types. This was the golden age of machine development; electronics was in its infancy, and the rotating machine was king. Many machines are now obsolete which were once made in large numbers. Thus the cross-field DC machines, or rotary amplifiers, have been replaced by solid-state power amplifiers; while the Schrage motor and other ingenious variable-speed AC machines have given way to the thyristor-controlled DC motor and the inverter-fed induction motor.

Two trends are discernible in the modern world of electrical machines. Firstly, there is a concentration on a small number of basic machine types which may be combined with power electronic controllers to form complete drive systems. A wide range of performance characteristics is easily achieved electronically, particularly with microprocessor-based systems. Secondly, new machines have evolved as other technologies have developed. Examples are superconducting machines, both AC and DC, linear motors and stepper motors. A particularly interesting development is the switched reluctance motor, described in chapter 5, in which the design of the machine and the electronic controller are closely linked.

Most of the machines in use today are derived from the three basic types treated in this book: simple DC machines, AC synchronous machines and AC induction machines. In spite of their superficial differences they all exploit the effects discovered by Oersted and Faraday. These two effects and their interrelation are seen most clearly when a current-carrying conductor is free to move in a magnetic field of constant intensity; this case is considered in section 1.2. But in most practical machines the conductors are not free to move; they are embedded in slots in the iron core of the machine, and forces act on the iron as well as on the conductors. It is necessary therefore to consider force production and EMF generation in more general terms, which is done in sections 1.3 and

1.4. Magnetic materials form an essential part of electrical machines; some of their properties are discussed in section 1.5, and the important concept of the magnetic circuit is introduced in section 1.6.

1.2 Conductor moving in a magnetic field

When a conductor moves in a magnetic field, an EMF is generated; when it carries a current in a magnetic field, a force is produced. Both of these effects may be deduced from one of the most fundamental principles of electromagnetism, and they provide the basis for a number of devices in which conductors move freely in a magnetic field. It has already been mentioned that most electrical machines employ a different form of construction, and the concepts developed in the next two sections are necessary for a proper understanding of their operation. Nevertheless the equations developed in this section for the force and the induced EMF remain valid for many practical machines; this important and useful result will be justified in chapter 2.

Induced EMF in a moving conductor

Consider a conductor moving with a velocity denoted by the vector $\boldsymbol{u}$ in a magnetic field $\boldsymbol{B}$ (figure 1.1). If the conductor slides along wires connected to a voltmeter, there will be a reading on the meter, showing that an EMF is being generated in the circuit. The effect may be explained in terms of the Lorentz equation for the force on a moving charge q

$$\boldsymbol{F} = q(\boldsymbol{E} + \boldsymbol{u} \times \boldsymbol{B}) \quad \text{newtons} \tag{1.1}$$

where q is the charge in coulombs, $\boldsymbol{E}$ the electric field strength in volts/metre, $\boldsymbol{u}$ the velocity in metres/second, and $\boldsymbol{B}$ the magnetic flux density in teslas. If the conductor is initially at rest, there will be no electric field $\boldsymbol{E}$ and no reading on

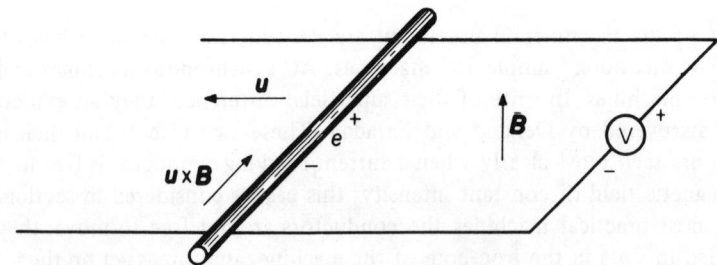

Figure 1.1 Moving conductor: induced EMF

the voltmeter. When the conductor is moving with velocity u, a force $qu \times B$ will act on any charged particle within the conductor, and the free charges (conduction electrons) will begin to move under the action of the force. There will be an accumulation of negative charge at one end of the conductor, leaving a surplus of positive charge at the other end; this will set up an electrostatic field E, and charge separation will continue until the force qE is exactly equal and opposite to $qu \times B$. The net force is then zero; there is no further motion of charge, and we have

$$E = -u \times B \tag{1.2}$$

The quantity $u \times B$ may be regarded as an induced electric field produced by the motion of the conductor, and this is opposed by an equal and opposite electrostatic field E produced by a distribution of electric charge. In virtue of the electrostatic field E there will be an electrostatic potential difference between the ends of the conductor given by the line integral of E along any path joining the ends PQ

$$v = -\int_P^Q E \cdot ds \text{ volts} \tag{1.3}$$

and this will be measured by a voltmeter connected between the wires. From eqn (1.2) this may be written as $v = e$, where

$$e = \int_P^Q u \times B \cdot ds \text{ volts} \tag{1.4}$$

and e may be regarded as the EMF induced in the conductor by its motion in the magnetic field. The integral may conveniently be taken along the axis of the conductor, which is a line of length l denoted by the vector l = PQ. If u, B and l are mutually perpendicular, the induced electric field $u \times B$ will be parallel to l, and its direction is determined by the right-hand screw rule of the vector product: if you look in the direction of $u \times B$, then B is displaced clockwise from u. For a magnetic field which is uniform along the length of the conductor, we then have

$$e = \int_0^l uB \, ds = Blu \text{ volts} \tag{1.5}$$

and the sign of e is determined from the direction of $u \times B$ as shown in figure 1.1. Equation (1.5) is commonly known as the 'flux cutting rule', or more accurately as the *motional induction formula*; its relation to Faraday's law of electromagnetic induction is discussed in section 1.3.

Conductor resistance

Suppose that a resistor is connected in place of the voltmeter, so that a current

i flows. If this current is distributed uniformly over the cross-section of the conductor, and the cross-sectional area is A, the magnitude of the current density is

$$J = \frac{i}{A} \text{ amperes/metre}^2 \tag{1.6}$$

and its direction is along the conductor, as shown in figure 1.2. Since the total force acting on unit charge is $E + u \times B$, Ohm's law for the moving conductor is

$$J = \sigma(E + u \times B) \tag{1.7}$$

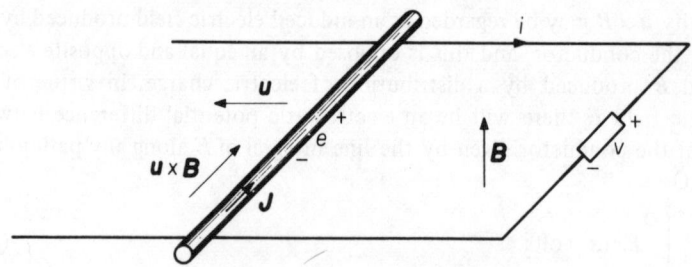

Figure 1.2 Moving conductor: induced current

where σ is the conductivity of the material. Thus the electrostatic field E must be slightly less than the induced electric field $u \times B$ when a current is flowing, the resultant force per unit charge being just sufficient to maintain the flow of current. The potential difference between the ends of the conductor is now given by

$$v = -\int E \cdot ds = \int u \times B \cdot ds - \int \frac{1}{\sigma} J \cdot ds$$

$$= \int_0^l uB \, ds - \int_0^l \frac{i}{\sigma A} \, ds = Blu - \frac{li}{\sigma A} \tag{1.8}$$

or

$$v = e - Ri \tag{1.9}$$

where $e = Blu$ is the induced EMF in volts and $R = l/\sigma A$ is the resistance of the conductor in ohms. The system may be represented by an equivalent circuit, as shown in figure 1.3.

Electromagnetic force on a conductor

Take the same configuration of a conductor in a magnetic field, and suppose initially that the conductor is stationary (figure 1.4). If a current i is flowing,

GENERAL PRINCIPLES

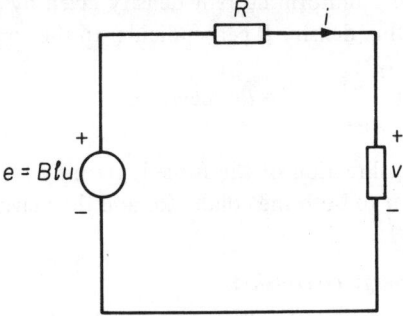

Figure 1.3 Equivalent circuit for a moving conductor

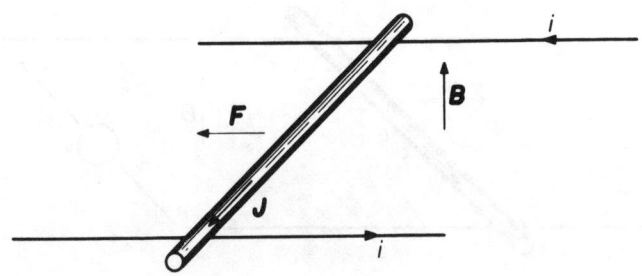

Figure 1.4 Force on a current-carrying conductor

there will be a flow of free charge along the conductor; let ρ be the charge per unit volume, and U the average drift velocity of the charge. This moving charge will experience a force in a magnetic field, and from eqn (1.1) the force per unit volume is

$$f = \rho U \times B \quad \text{newtons/metre}^3 \qquad (1.10)$$

Since the free charge cannot escape from the sides, this force is transmitted to the conductor. We may express the force in terms of the current by noting that the current density J is given by

$$J = \rho U \qquad (1.11)$$

The force per unit volume is therefore

$$f = J \times B \qquad (1.12)$$

and the total force on the conductor is given by the volume integral

$$F = \int J \times B \, dv \quad \text{newtons} \qquad (1.13)$$

In the simple case of a uniform current density given by $J = i/A$ (eqn 1.6), and a uniform magnetic flux density B perpendicular to the conductor

$$F = \int JB \, dv = \int_0^l \frac{iB}{A} A \, ds = Bli \text{ newtons} \tag{1.14}$$

From eqn (1.12) the direction of the force is given by the vector product $\boldsymbol{J} \times \boldsymbol{B}$, and it is perpendicular to both the conductor and the magnetic field.

Electromechanical energy conversion

Figure 1.5 shows the conductor connected to a voltage source v. There is a current i flowing, and the conductor is moving with a velocity $\boldsymbol{u}$. The directions

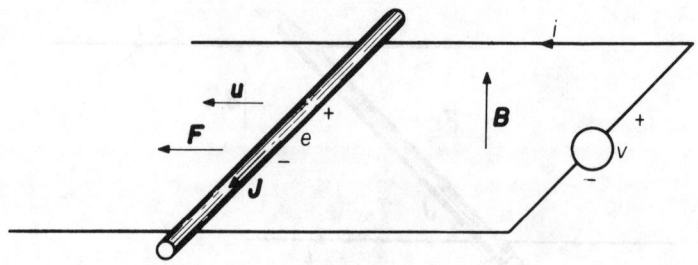

Figure 1.5   Moving current-carrying conductor

of the force $\boldsymbol{F}$ and the induced EMF e are shown in the figure. Since the force $\boldsymbol{F}$ and the velocity $\boldsymbol{u}$ are in the same direction, the conductor does mechanical work at the rate

$$P_m = \boldsymbol{F} \cdot \boldsymbol{u} = Fu \text{ watts} \tag{1.15}$$

The voltage source v is driving a current i into the circuit, and it therefore does work at the rate

$$P_e = vi \text{ watts} \tag{1.16}$$

Since the direction of current flow is the reverse of that in figure 1.2, eqn (1.9) becomes

$$v = e + Ri$$
$$= Blu + Ri \tag{1.17}$$

and we also have the force equation

$$F = Bli \tag{1.14}$$

Multiplying eqn (1.17) by i and eqn (1.14) by u gives

$$P_e = vi = Blui + Ri^2$$

$$P_m = Fu = Bliu$$

It follows that

$$P_e = P_m + Ri^2 \qquad (1.18)$$

showing that the electrical input power P_e is equal to the mechanical output power P_m plus the ohmic losses in the conductors; the device is acting as a motor, converting electrical energy into mechanical energy. For the current to flow in the direction shown, the applied voltage v must exceed the induced EMF e; if v is smaller than e the direction of current flow is reversed, and the direction of the force F is reversed in consequence. The conductor then absorbs mechanical energy at the rate $P_m = Fu$; the voltage source likewise absorbs electrical energy at the rate $P_e = vi$, and

$$P_m = P_e + Ri^2 \qquad (1.19)$$

The device is acting as a generator, converting mechanical energy into electrical energy plus ohmic losses. Thus the process of energy conversion is reversible, and there is no fundamental difference between generator and motor action.

Applications

Among the best-known applications of the conductor in a magnetic field are the moving-coil loudspeaker (figure 1.6) and the moving-coil meter (figure 1.7). In each of these devices the conductor, in the form of a coil, moves in the uniform radial field of a permanent magnet; motion of the coil is opposed by a spring, giving a displacement proportional to the coil current. The direct proportionality of the force or displacement to the current makes the moving-coil principle particularly useful for instrumentation. In the force-balance accelerometer, for instance, a feedback system senses the coil displacement and adjusts the current until the electromagnetic force exactly balances the acceleration reaction force. The coil current then gives a measure of the acceleration precise enough for modern inertial navigation techniques.

Another application is a special type of electrical machine. Most conventional motors and generators depend for their operation on the force between magnetised iron parts; but in the homopolar machine the force is developed directly on a conductor moving in a magnetic field of constant intensity. Faraday in 1831 made the first generator using this principle, in the form of a circular disc rotating in a magnetic field (figure 1.8). Each element of the disc at a radial distance r from the axis is moving with velocity $u = \omega r$ perpendicular to the magnetic field. The induced electric field $\boldsymbol{u} \times \boldsymbol{B}$ is directed along the radius, so there will be an

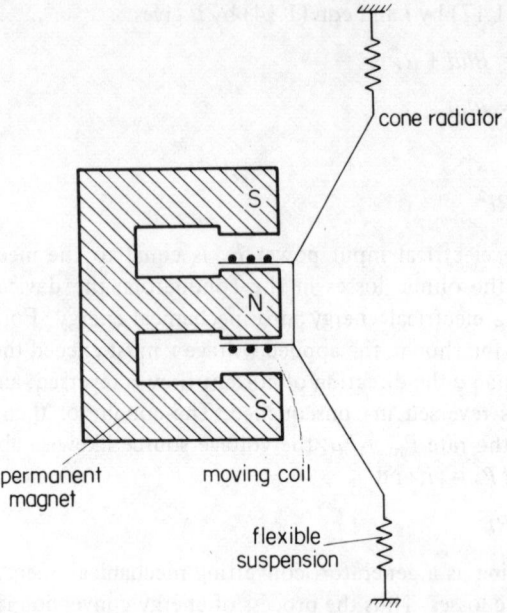

Figure 1.6 Moving-coil loudspeaker

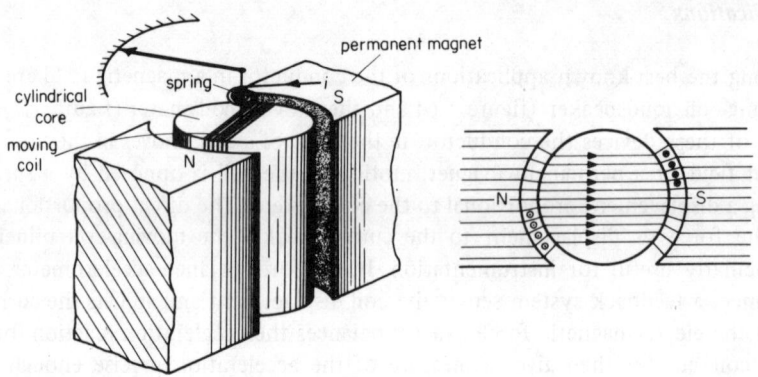

Figure 1.7 Moving-coil meter

EMF induced between the centre of the disc and the periphery. Integration along a path such as PQ gives

$$e = \int_P^Q \boldsymbol{u} \times \boldsymbol{B} \cdot \mathrm{d}\boldsymbol{s} = \int_0^a B\omega r\, \mathrm{d}r = \tfrac{1}{2}B\omega a^2 \quad \text{volts} \tag{1.20}$$

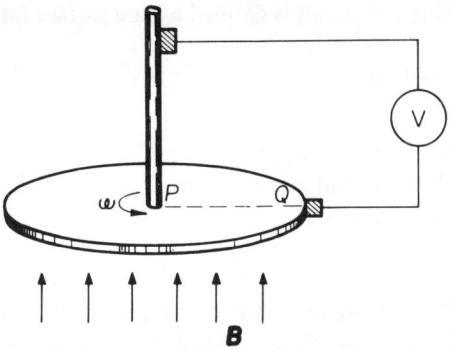

Figure 1.8 Faraday disc generator

where a is the radius of the disc in metres and ω the angular velocity in radians/second. The generated voltage is rather low for small machines at normal speeds, if the flux density is limited (as it usually is) by the saturation of iron to about 2 teslas. For example, if $a = 100$ mm, $\omega = 100$ rad/s (≈ 1000 rev/min) and $B = 2$ T, then $e = 1$ V. Large machines have been built for special low-voltage heavy-current applications, and an interesting development is the use of superconducting coils to generate very high magnetic fields [2]. Superconducting machines can be smaller, lighter and more efficient than their conventional counterparts, but their commercial application is restricted at present by the poor reliability of liquid helium refrigerators for the superconducting coils.

1.3 Electromagnetic induction

It was shown in section 1.2 that an EMF is induced in a conductor when it moves in a magnetic field. An EMF can also be induced in a stationary circuit by a time-varying magnetic field. If this magnetic field is produced by currents flowing in conductors or coils, the EMF can be induced merely by changing the current; no motion of any part of the system is required. The effect is termed transformer induction, and it appears to be physically quite distinct from motional induction. Both effects are included in Faraday's law of electromagnetic induction, which relates the induced EMF in a circuit to the rate of change of the magnetic flux linking the circuit.

Flux linkage

If a circuit consists of a conductor in the form of a simple closed curve C, the magnetic flux Φ linking the circuit is defined by the surface integral

$$\Phi = \int_S \boldsymbol{B} \cdot \mathrm{d}\boldsymbol{a} \text{ webers} \tag{1.21}$$

where S is any surface spanning the boundary C of the circuit. If the magnetic field is uniform and the circuit has an area A perpendicular to the field, this reduces to the simple expression

$$\Phi = BA \tag{1.22}$$

The concept of flux linkage arises when it is desired to calculate the flux linking a multi-turn coil. It is possible in principle to devise a twisted surface resembling an Archimedian screw, bounded by the turns of the coil, and to evaluate the integral in eqn (1.21) over this surface. But it is simpler to suppose that each turn links a certain amount of flux, so that the total flux linking the coil is the sum of the contributions from the individual turns. Thus if each turn links a flux Φ and the coil has N turns, then the total flux linking the coil, or flux linkage, is given by

$$\psi = N\Phi \text{ webers} \tag{1.23}$$

If the magnetic field is uniform and parallel to the axis of the coil, and each turn has an area A, then

$$\psi = NBA \tag{1.24}$$

Usually the field is not uniform and the flux through an individual turn will depend on its position (figure 1.9). The total flux linking the coil is then given by the sum

$$\psi = \sum_{r=1}^{N} \Phi_r \text{ webers} \tag{1.25}$$

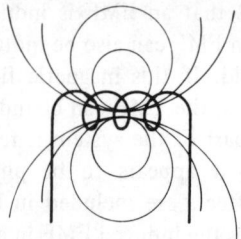

Figure 1.9 Flux linkage: non–uniform field

and an average flux per turn may be defined by the relation

$$\Phi_{av} = \psi/N \tag{1.26}$$

Inductance

If permanent magnets are excluded, the flux linking a coil will depend on (a) the current flowing in the coil and (b) currents flowing in any adjacent coils or conductors.

Self-inductance

With a single coil carrying a current i we have

$$\psi = f(i) \tag{1.27}$$

When there are no ferromagnetic materials present (the coil is air-cored) the relationship is linear; thus

$$\psi = Li \tag{1.28}$$

where L is a constant known as the *self-inductance* of the coil. The unit of L is the henry when ψ is in webers and i in amperes. When the coil has an iron core the relationship between ψ and i is no longer linear, on account of the magnetic properties of the iron. The form of eqn (1.28) may still be used, but the coefficient L is no longer a constant; this can be made explicit by writing

$$\psi = iL(i) \tag{1.29}$$

In order to simplify the analysis it is often assumed that the inductance of an iron-cored coil is a constant; this assumption must be used with caution, for it can sometimes give completely erroneous results. This point will be discussed more fully in section 1.6.

Mutual inductance

With two coils the flux linkages are functions of the coil currents and the geometry of the system. Thus

$$\begin{aligned} \psi_1 &= L_1 i_1 + M_{12} i_2 \\ \psi_2 &= L_2 i_2 + M_{21} i_1 \end{aligned} \tag{1.30}$$

where ψ_1 and ψ_2 are the flux linkages for the two coils, and i_1 and i_2 are the corresponding currents. The coefficients M_{12} and M_{21} are known as the *mutual inductances*. When the coils do not have iron cores, it may be shown [3] that the mutual inductance coefficients are constant and equal, that is

$$M_{12} = M_{21} = M \tag{1.31}$$

If ψ_{12} is the flux linking the first coil due to a current i_2 in the second, and ψ_{21} is the flux linking the second coil due to a current i_1 in the first, then the mutual inductance is given by

$$M = \frac{\psi_{12}}{i_2} = \frac{\psi_{21}}{i_1} \tag{1.32}$$

This reciprocal property is particularly useful when the mutual inductance has to be measured or calculated, for one of the two alternative expressions in eqn (1.32) is often easier to evaluate than the other.

Faraday's law

Faraday's law of electromagnetic induction states that the EMF induced in a circuit is proportional to the rate of change of flux linkages. In SI units the constant of proportionality is unity, so that

$$e = \pm \frac{d\psi}{dt} \text{ volts} \tag{1.33}$$

The question of the sign in eqn (1.33) sometimes causes difficulty. Traditionally a negative sign is used in deference to Lenz's law, which states that any current produced by the EMF tends to oppose the flux change. But this is inconsistent with the definition of inductance given in eqn (1.28) and the usual circuit conventions shown in figure 1.10. If we take the positive sign in eqn (1.33) and substitute for ψ from eqn (1.28), then

$$e = +L \frac{di}{dt} \tag{1.34}$$

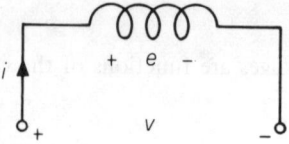

Figure 1.10 Induced EMF in a coil

For a pure inductance, with no internal resistance, Kirchhoff's voltage law applied to the circuit gives

$$v - e = 0$$

so that

$$v = e \tag{1.35}$$

Consequently the terminal voltage v is given by

$$v = +L \frac{di}{dt} \tag{1.36}$$

and this is the equation which defines the inductance element in circuit theory. The correct form of Faraday's law is therefore

$$e = +\frac{d\psi}{dt} \tag{1.37}$$

and Lenz's law may be used to resolve any uncertainty about the positive directions of e and ψ in the circuit.

Calculation of the induced EMF

Faraday's law relates the induced EMF to the rate of change of flux linkage, regardless of the way in which the change occurs. The flux linkage of a circuit may be changed in several ways: the strength of the magnetic field may be altered, either by moving the circuit relative to the source of the field or by varying the currents which create the field, or the boundary of the circuit may be deformed while the magnetic field remains unchanged. The moving conductor in section 1.2 is an example of this last case. Consider the circuit formed by the voltmeter, the fixed rails, and the moving conductor; the area of this circuit increases steadily, and the rate of change of flux is equal to *Blu*. This result agrees with the previous calculation. In all cases the induced EMF may be calculated by the direct application of Faraday's law, and this is the only satisfactory method when motional and transformer effects are both present.

Particular care is needed when calculating the EMF in a moving conductor. It is tempting to use the 'flux cutting formula' $e = Blu$ in all cases, but this can give incorrect results when parts of the magnetic structure move with the conductor. The derivation of the formula given in section 1.2 is for the particular case of a conductor whose motion does not affect the source of the magnetic field in any way, and its direct application is limited to that situation. More complex problems can be treated by expressing the total magnetic field as the sum of components from different parts of the magnetic structure, and taking the sum of *Blu* terms with the appropriate values of u [4]. But the direct application of Faraday's law is the safest procedure in this kind of problem. Carter [3] gives a particularly good discussion of electromagnetic induction and some apparent paradoxes.

Induced EMF and inductance

The self-inductance and mutual-inductance coefficients can often be changed by relative movement of parts of the system, and Faraday's law gives the correct value for the induced EMF in these cases. For example, with the single coil shown in figure 1.10 the induced EMF is given by

$$e = \frac{d\psi}{dt} = \frac{d}{dt}(Li) = L\frac{di}{dt} + i\frac{dL}{dt} \qquad (1.38)$$

Thus if the motion of a part of the system causes L to change, there will be an EMF term additional to the normal EMF of self-induction. The voltage equation for the circuit should therefore be written as

$$v = Ri + \frac{d}{dt}(Li) \qquad (1.39)$$

Similarly, with the coupled coils shown in figure 1.11 the voltage equations are

$$v_1 = R_1 i_1 + \frac{d}{dt}(L_1 i_1 + M i_2)$$
$$v_2 = R_2 i_2 + \frac{d}{dt}(L_2 i_2 + M i_1) \qquad (1.40)$$

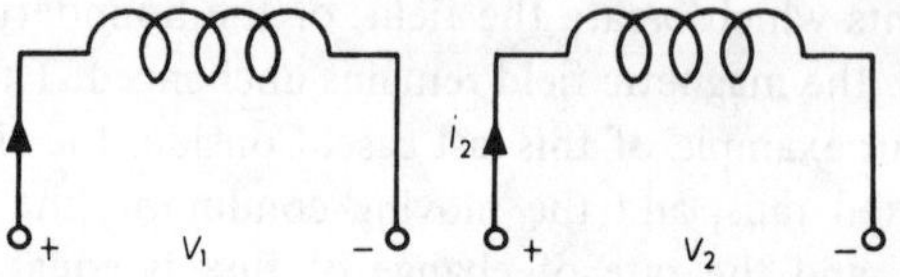

Figure 1.11 Coupled coils

When there is no motion of parts of the system the inductance coefficients are constant, and these equations reduce to the ordinary equations of coupled circuit theory.

1.4 Electromagnetic forces

In section 1.2 a method was given for calculating the force on a conductor in a magnetic field. In many practical devices, including rotating machines, magnetic forces act on the iron parts as well as on the conductors. These forces on the magnetised iron parts are often the dominant ones, and there is a need to calculate the total electromagnetic force acting on a structure made up of conductors and ferromagnetic materials. Two methods of calculation are given in this section. The first is the Maxwell stress method, which also provides a useful physical picture of the mechanism of force production. The second is an energy method, which complements the Maxwell stress method for purposes of calculations, but is less useful as a physical explanation.

The Maxwell stress concept

There is a sound scientific basis to the elementary idea that the magnetic lines of force are like rubber bands tending to draw pieces of iron together. The idea of lines or tubes of force was central to Faraday's conception of the magnetic field, but it was Maxwell who gave precise mathematical expression to this concept. Maxwell showed, as a deduction from the equations of the electromagnetic field, that magnetic forces could be considered to be transmitted through space (or a non-magnetic material) by the following system of stresses [3, 5]

(a) a tensile stress of magnitude $\frac{1}{2}BH$ newtons per square metre along the lines of force
(b) a compressive stress, also of magnitude $\frac{1}{2}BH$ newtons per square metre, at right angles to the lines of force.

Since $B = \mu_0 H$ in a non-magnetic medium, the stresses may also be written as $\frac{1}{2}\mu_0 H^2$ or $B^2/2\mu_0$.

If the magnetic field is perpendicular to the surface of a body (figure 1.12), there will be a tensile stress of magnitude $B^2/2\mu_0$, also perpendicular to the surface, drawing the surface into the field. If the field is parallel to the surface (figure 1.13), there will be a compressive stress of magnitude $B^2/2\mu_0$ pushing the surface out of the field. In the general case, when the flux density B makes an angle θ with the normal n, the stress t makes an angle 2θ with n (figure 1.14). The magnitude of t is still $B^2/2\mu_0$, and the three vectors n, B and t are coplanar.

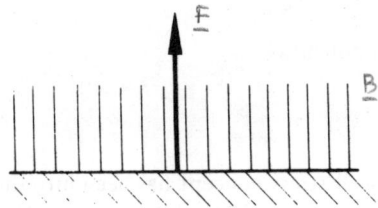

Figure 1.12   Magnetic force: field perpendicular to the surface

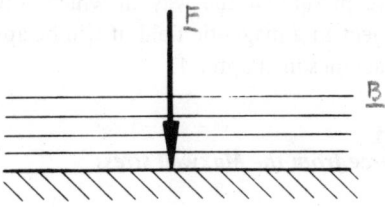

Figure 1.13   Magnetic force: field parallel to the surface

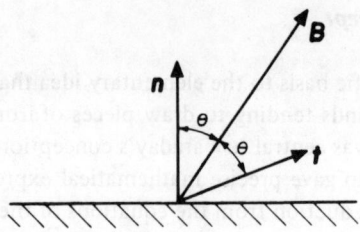

Figure 1.14 Maxwell stress vector

The force on an element of area δA is thus in the direction of t, and its magnitude is given by

$$\delta F = t\, \delta A = \frac{B^2}{2\mu_0}\, \delta A \quad \text{newtons} \tag{1.41}$$

An interesting, and at first sight surprising, deduction is that the force will be parallel to the surface when the field is inclined at 45°.

It is sometimes useful to relate the components of stress to the components of flux density, which may be done by resolving the vectors in directions normal and tangential to the surface. Thus if B_n and B_s are respectively the normal and tangential components of B, the normal component of stress is given by

$$t_n = \frac{1}{2\mu_0}(B_n^2 - B_s^2) \tag{1.42}$$

and the tangential component is

$$t_s = \frac{B_n B_s}{\mu_0} \tag{1.43}$$

Although the Maxwell stress concept has been introduced in terms of magnetised iron parts, it is not restricted to this situation. The electromagnetic force acting on any combination of iron parts and conductors may be found from the Maxwell stress on a surface enclosing the bodies; the only restriction is that the surface should not pass through any magnetised parts. The concept gives an immediate qualitative picture of the way in which forces are distributed over the surface of an object in a magnetic field; it will be applied to DC machines in chapter 2 and AC machines in chapter 4.

Calculation of the force from the Maxwell stress

Equations (1.42) and (1.43) may be integrated over the surface to give the total force on an object; this presupposes that an accurate field solution is available, for example by numerical analysis [6]. When such a solution is not available, it is still possible to calculate the force approximately from the Maxwell stress. Two examples will illustrate the method.

Force of attraction

Figure 1.15 shows an electromagnet lifting an iron bar, and the problem is to calculate the force of attraction. The actual field pattern is quite complex; it would be very tedious first to solve the field equations with the boundary con-

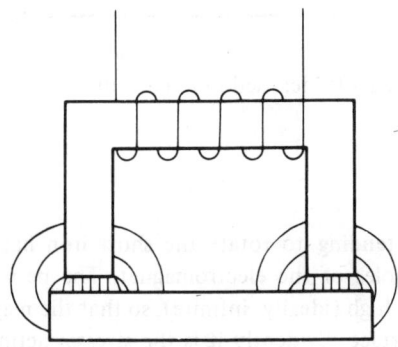

Figure 1.15 Electromagnet: actual field pattern

ditions imposed by the structure, and then to integrate the stress over the surface to find the force. A good approximation can be obtained from the following considerations. If the airgaps between the magnet poles and the bar are small, and the permeability of the magnetic material is high, the magnetic field in the gaps will be nearly uniform and much more intense than the fringing field outside the airgaps. (This will be justified formally in section 1.6.) Since the Maxwell stress varies as B^2, the field outside the airgaps will contribute very little to the total force, and it may be ignored. For the purpose of calculating the force, we may therefore replace the actual field distribution of figure 1.15 with the idealised distribution of figure 1.16, in which the field is uniform, confined to the airgaps and normal to the iron surfaces. If the area of each pole face is A square metres, and the magnetic flux density in each airgap is B teslas, the total force is

$$F = \frac{B^2}{2\mu_0} \cdot 2A = \frac{AB^2}{\mu_0} \text{ newtons} \qquad (1.44)$$

A method of calculating the flux density from the coil current and the dimensions of the magnet will be given in section 1.6.

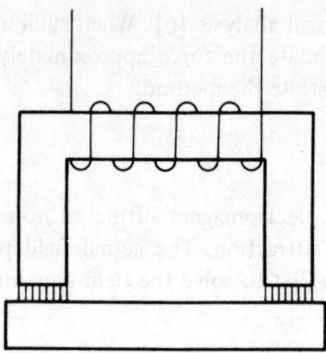

Figure 1.16 Electromagnet: idealised field pattern

Alignment torque

Consider the torque tending to rotate the short iron bar in figure 1.17 into alignment with the poles of the electromagnet. The permeability of the iron is assumed to be very high (ideally, infinite), so that the magnetic field is always normal to the iron surface. Evidently it is the stresses acting on the portions X and Y of the iron surface that are tending to rotate the bar; but the field here is particularly difficult to calculate.

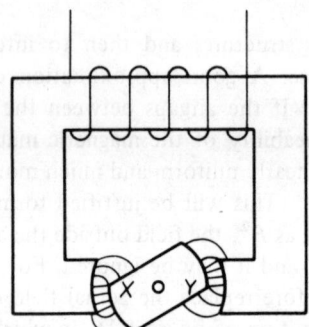

Figure 1.17 Alignment torque on an iron rotor

This is an example of a problem which requires another analytical device. The integral of the Maxwell stress over any surface surrounding a magnetic object gives the correct value for the total force on the object, even though the distribution of the force over this surface is quite different from the distribution over the surface of the object. To apply this to the alignment problem, observe that the field in the narrow airgap is practically uniform and can be calculated by the methods of section 1.6. Choose a surface such as JKLM (figure 1.18), and ignore the field beyond the corners E, F, G and H. The portions EJ, FL, KG and MH are perpendicular to the field; there will be tensile forces on these surfaces which

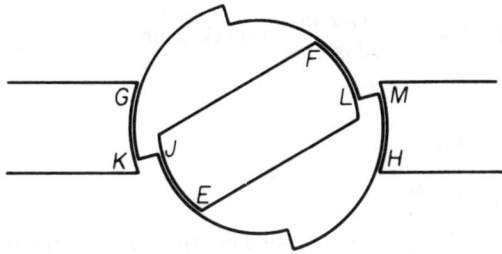

Figure 1.18 Surface for Maxwell stress calculation

cancel out in pairs. Since the field is negligible beyond the corners, there is no force on the portions FG and EH. The portions JK and LM are both parallel to the field; there will be compressive stresses on these surfaces tending to rotate the bar in a clockwise direction. If JK = LM = g and the depth of the bar is d, the force on each surface is

$$F = \frac{B^2}{2\mu_0} \cdot gd \text{ newtons}$$

and if a is the mean distance from the pivot to the airgap, the torque is

$$T = 2Fa = adgB^2/\mu_0 \text{ newton metres} \tag{1.45}$$

Energy methods

Work must be done to establish a magnetic field, and the energy stored in the field is given by

$$W_m = \int \tfrac{1}{2}BH \, dv \text{ joules} \tag{1.46}$$

where the integral is taken over the whole volume of the field.

Consider a system consisting of pieces of magnetic material together with coils or conductors carrying currents. In general there will be electromagnetic forces acting on the various parts of the system, and if any part is displaced the force will do work. Let there be a small displacement δx in some part of the system. If the component of force in the direction of the displacement is F_x, the work done will be

$$\delta W = F_x \, \delta x \qquad (1.47)$$

During this displacement there may be an increase δW_m in the stored magnetic energy, and if voltages are induced in any of the coils the electrical sources will have to supply an amount of energy δW_e. We thus have

energy supplied = increase in stored energy + work done

so that

$$\delta W_e = \delta W_m + \delta W$$
$$= \delta W_m + F_x \, \delta x \qquad (1.48)$$

If the currents in the coils are adjusted continuously during the displacement so that there is no change in the flux linkages, there will be no induced voltages; consequently the energy supplied, δW_e, will be zero. The work done by the force must come from the energy stored in the field

$$F_x \, \delta x = -\delta W_m$$

so that

$$F_x = -\frac{\partial W_m}{\partial x}\bigg|_{\text{constant flux}} \qquad (1.49)$$

When there is a linear relationship between flux and current another expression may be obtained. If the currents in all the coils are held constant during the displacement, it may be shown [3] that the energy δW_e supplied by the sources is equally divided between the mechanical work $F_x \, \delta x$ and the increase in stored energy δW_m. Thus

$$F_x \, \delta x = \delta W_m$$

so that

$$F_x = +\frac{\partial W_m}{\partial x}\bigg|_{\text{constant current}} \qquad (1.50)$$

The force will be in newtons when the displacement is in metres and the field energy is in joules. Similar equations hold for rotational motion if F_x is replaced by the torque T_θ (in newton metres) and x is replaced by the angular displacement θ (in radians).

Calculation of the force on an iron part

As an example of the use of these expressions, consider once again the electromagnet shown in figure 1.16. If x is the displacement of the bar from the poles, the field energy is given by

$$W_m = \int_{\text{airgap}} \tfrac{1}{2} BH \, dv + \int_{\text{core}} \tfrac{1}{2} BH \, dv$$

$$= \frac{B^2 A x}{\mu_0} + \int_{\text{core}} \tfrac{1}{2} BH \, dv \text{ joules} \qquad (1.51)$$

If the flux linkage is constant, B will be constant and the energy stored in the core will also be constant. Therefore

$$F_x = -\left.\frac{\partial W_m}{\partial x}\right|_{\text{constant flux}} = -\frac{B^2 A}{\mu_0} \text{ newtons} \qquad (1.52)$$

This is numerically the same as eqn (1.44) obtained from the Maxwell stress, and the negative sign shows that the force on the bar is in the direction of decreasing x, that is, upwards. Here the Maxwell stress method is obviously simpler, and Carpenter [7] has shown this to be true generally for calculating forces on iron surfaces.

Energy and inductance

With a single coil carrying a current i, the energy stored in the magnetic field is given by

$$W_m = \tfrac{1}{2} L i^2 \qquad (1.53)$$

If the motion of a part of the system causes a change in the inductance, then eqn (1.50) gives

$$F_x = \left.\frac{\partial W_m}{\partial x}\right|_{\text{constant current}} = \tfrac{1}{2} i^2 \frac{\partial L}{\partial x} \qquad (1.54)$$

This is an important and useful result, for it shows that the mechanical force can be expressed in terms of the variation in the inductance coefficient L, a quantity which can be measured electrically.

With a pair of mutually coupled coils carrying currents i_1 and i_2, the stored magnetic energy is given by

$$W_m = \tfrac{1}{2} L_1 i_1^2 + \tfrac{1}{2} L_2 i_2^2 + M i_1 i_2 \qquad (1.55)$$

The force acting on a part of the system is then given by

$$F_x = \tfrac{1}{2} i_1^2 \frac{\partial L_1}{\partial x} + \tfrac{1}{2} i_2^2 \frac{\partial L_2}{\partial x} + i_1 i_2 \frac{\partial M}{\partial x} \qquad (1.56)$$

Similar expressions hold for torque in terms of angular displacement.

Calculation of the torque on an air-cored coil

As an example of the application of eqn (1.56), consider the coil system of an electrodynamic wattmeter, shown in figure 1.19. A small moving coil is mounted on pivots midway between two fixed coils. The fixed coils are connected in

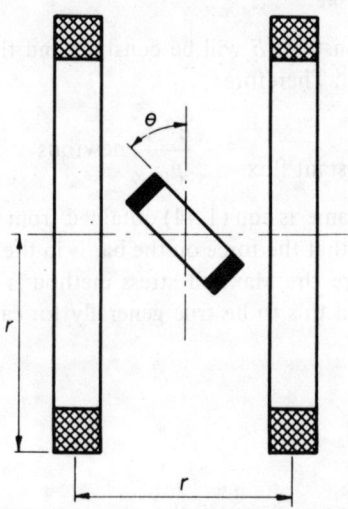

Figure 1.19 Coil system for an electrodynamic wattmeter

series, and are separated by a distance equal to their radius; this is a Helmholtz pair, which produces in the vicinity of the moving coil a nearly uniform magnetic field parallel to the common axis of the fixed coils. If each fixed coil has N_1 turns and carries a current i_1 amperes, the flux density is

$$B = \frac{8\mu_0 N_1 i_1}{5\sqrt{5}r} \text{ teslas} \tag{1.57}$$

where r is the radius in metres.

If the moving coil has N_2 turns and an area A square metres, the flux linking it is

$$\psi_{21} = N_2 AB \cos\theta \text{ webers} \tag{1.58}$$

and the mutual inductance between the fixed and moving coils is

$$M = \frac{\psi_{21}}{i_1} = \frac{8\mu_0}{5\sqrt{5}} \cdot \frac{N_1 N_2 A}{r} \cos\theta \text{ henrys} \tag{1.59}$$

Since the self-inductances are independent of θ, the torque on the moving coil is

$$T = i_1 i_2 \frac{\partial M}{\partial \theta} = -\frac{8\mu_0}{5\sqrt{5}} \cdot \frac{N_1 N_2 A}{r} i_1 i_2 \sin\theta \text{ newton metres} \qquad (1.60)$$

This is an instance of an energy method giving a straightforward calculation of the torque, whereas the Maxwell stress would be difficult to evaluate.

1.5 Magnetic materials

In free space the magnetic flux density B is related to the magnetising force H by the expression

$$B = \mu_0 H \qquad (1.61)$$

where μ_0 is the primary magnetic constant (with a value of $4\pi \times 10^{-7}$ H/m). This relationship is modified in a material medium; if we exclude permanent-magnet materials, eqn (1.61) becomes

$$B = \mu_0 \mu_r H \qquad (1.62)$$

where μ_r is a dimensionless number known as the relative permeability of the material. For most engineering applications of magnetism, materials may be divided into three groups: (a) permanent-magnet materials, for which eqn (1.62) does not hold; (b) ferromagnetic materials, typified by iron, for which the relative permeability μ_r is large and variable; (c) all other materials, for which μ_r is practically equal to unity. There is normally no need to consider the small deviations from unity which characterise paramagnetic and diamagnetic behaviour. Only a brief introduction to the properties of ferromagnetic materials will be given in this section; further information is available in standard texts such as Brailsford [8]. Permanent magnets are discussed in section 1.7.

A typical ferromagnetic material is silicon steel, which is widely used for the cores of transformers and rotating machines. When such a material is magnetised by slowly increasing the applied magnetising force H, the resulting flux density B follows a curve of the form shown in figure 1.20. This is known as the *magnetisation curve* for the material, and the corresponding variation of μ_r with B is shown in figure 1.21. If the magnetising force is gradually reduced to zero, the flux density does not follow the same curve; and if the magnetising force slowly alternates between positive and negative values, the relationship between B and H takes the form of a hysteresis loop as shown in figure 1.22. When the amplitude of the alternating magnetising force is changed, a new hysteresis loop will be formed; the locus of the tips of these loops is the magnetisation curve shown in figure 1.20.

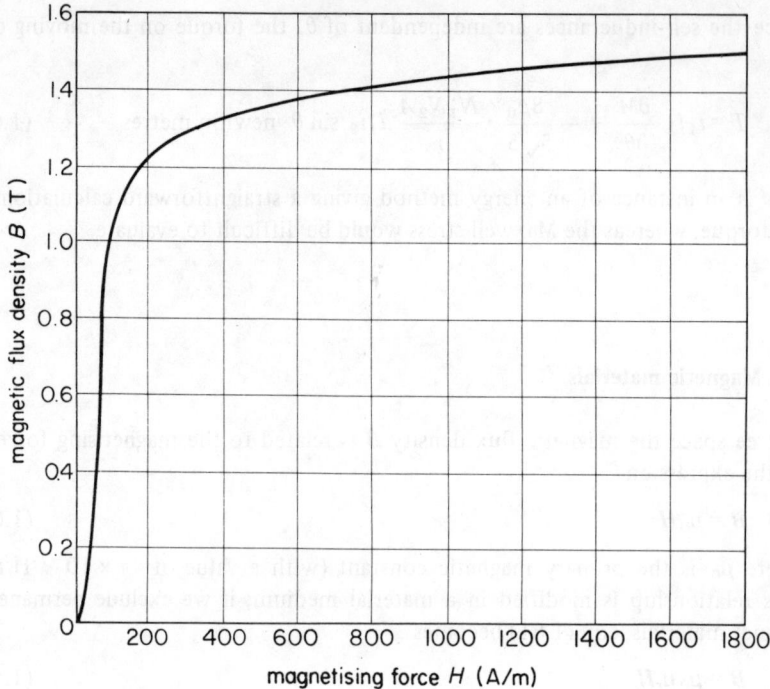

Figure 1.20 Magnetisation curve for 4 per cent silicon steel

The part of the magnetisation curve where the slope begins to change rapidly is termed the *knee*. Below the knee it is often possible to use a linear approximation to the actual characteristic, with a corresponding constant value for the relative permeability. But the onset of saturation above the knee marks a dramatic change in the properties of the material, which must be recognised in the design and analysis of magnetic structures.

Hysteresis loss

When a magnetic material is taken through a cycle of magnetisation, energy is dissipated in the material in the form of heat. This is known as the *hysteresis loss*, and it may be shown [3] that the energy loss per unit volume for each cycle of magnetisation is equal to the area of the hysteresis loop. The area of the loop will depend on the nature of the material and the value of B_{max} (figure 1.22), and an approximate empirical relationship discovered by Steinmetz is

$$w_h = \lambda_h B_{max}^n \text{ joules/metre}^3 \qquad (1.63)$$

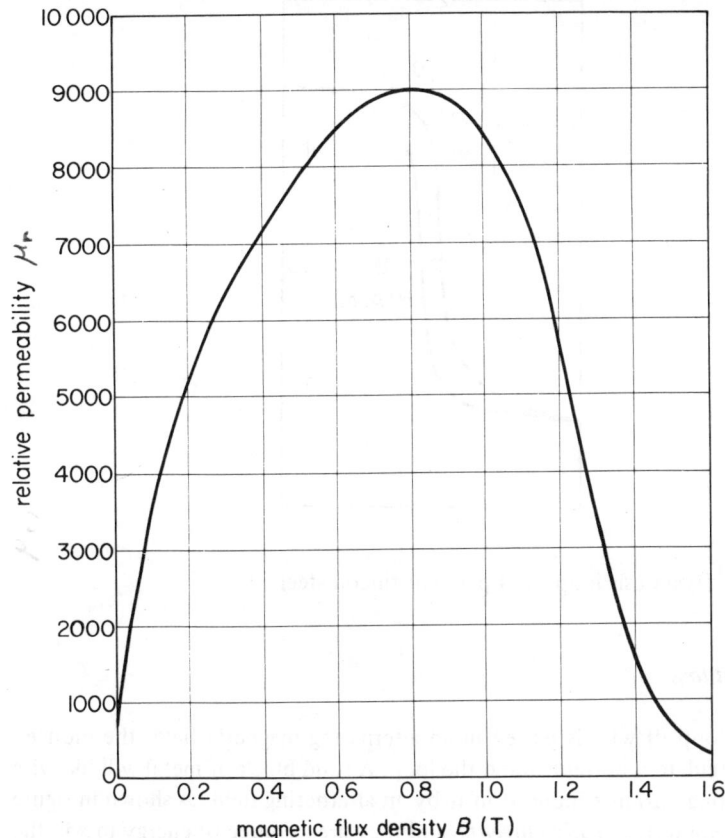

Figure 1.21 Relative permeability of 4 per cent silicon steel

In this expression w_h is the loss per unit volume for each cycle of magnetisation; the index n has a value of about 1.7 for many materials; and the coefficient λ_h is a property of the material, with typical values of 500 for 4 per cent silicon steel and 3000 for cast iron.

When the material is subjected to an alternating magnetic field of constant amplitude there will be a constant energy loss per cycle, and the power absorbed is therefore proportional to the frequency. Assuming the Steinmetz law, we have the following expression for the hysteresis loss per unit volume

$$p_h = \lambda_h B_{max}^{1.7} f \text{ watts/metre}^3 \tag{1.64}$$

where f is the frequency in hertz.

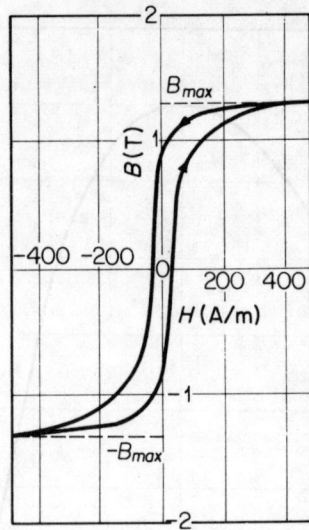

Figure 1.22 Hysteresis loop for 4 per cent silicon steel

Eddy current loss

If a closed loop of wire is placed in an alternating magnetic field, the induced EMF will circulate a current round the loop. A solid block of metal will likewise have circulating currents induced in it by an alternating field, as shown in figure 1.23. These are termed *eddy currents*, and they are a source of energy loss in the metal. Eddy current losses occur whenever conducting material is placed in a changing magnetic field; the magnitude of the loss is dependent on the properties of the material, its dimensions and the frequency of the alternating field.

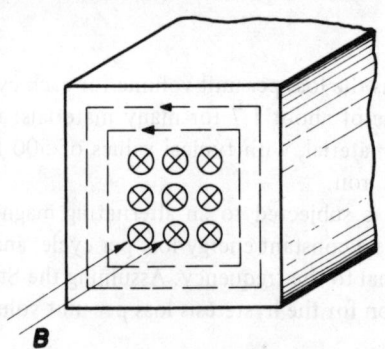

Figure 1.23 Eddy currents in a solid conductor

GENERAL PRINCIPLES

Magnetic structures carrying alternating magnetic flux are usually made from a stack of thin plates or laminations, separated from one another by a layer of insulation (figure 12.4). This construction breaks up the eddy current paths, with a consequent reduction in the loss; qualitatively, the effect may be explained as follows. With solid metal (figure 1.23) the currents would flow in approximately square paths; these paths enclose a large area for a given perimeter, and the induced EMF is high for a path of given resistance. When the metal is divided into laminations (figure 1.24), the current paths are long narrow rectangles; the area enclosed by a given perimeter is much smaller, and the induced EMF is smaller, giving lower currents and reduced losses. An approximate analysis [3, 8] shows that in plates of thickness t (where t is much smaller than the width or length) the eddy current loss per unit volume is given by

$$p_e = \frac{\pi^2 B_{max}^2 f^2 t^2}{6\rho} \text{ watts/metre}^3 \tag{1.65}$$

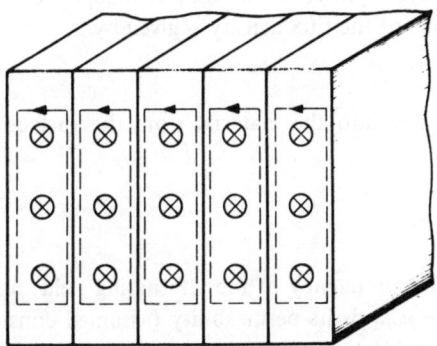

Figure 1.24 Eddy currents in a laminated conductor

where the flux density is an alternating quantity of the form

$$B = B_{max} \sin 2\pi f t \tag{1.66}$$

and ρ is the resistivity of the material. Thus if the lamination thickness is reduced by a factor x, the loss is reduced by a factor x^2. As might be expected, the loss varies inversely with the resistivity ρ. The addition of 3-4 per cent of silicon to iron increases the resistivity by about four times, as well as reducing the hysteresis loss; this is the main reason for the widespread use of silicon steel in electrical machines. The thickness of the laminations is typically 0.3-0.5 mm, which ensures that the eddy current loss will be less than the hysteresis loss at a frequency of 50 Hz.

Skin effect

The eddy currents in a bar such as the one shown in figure 1.23 will produce a magnetic field within the bar which, by Lenz's law, will oppose the applied field. Thus the magnetic flux density will fall from a value B_0 at the surface to some lower value in the interior. The effect depends on the properties of the material, the frequency of the alternating field and the dimensions of the bar. It is possible for the magnitude of the flux density to fall very rapidly in the interior of the bar, so that most of the flux is confined to a thin layer or skin near the surface. The phenomenon is termed *skin effect*, and it implies very inefficient use of the magnetic material (quite apart from any eddy current losses). A similar effect occurs in conductors carrying alternating current, where the current density falls from some value J_0 at the surface to a lower value in the interior.

The variation of flux density with distance may be calculated by solving the electromagnetic field equations [9]. When skin effect is well developed, so that the flux density decays rapidly, the solution is independent of the geometry of the bar; the magnitude of the flux density is given by

$$B = B_0 e^{-x/\delta} \qquad (1.67)$$

where x is the distance into the material from the surface. The quantity δ is given by

$$\delta = \sqrt{\frac{2\rho}{\mu\omega}} \qquad (1.68)$$

where ω is the angular frequency of the alternating field, ρ is the resistivity of the material and $\mu = \mu_0 \mu_r$ is its permeability (assumed constant). At a depth δ the magnitude of B is $1/e$ times the value at the surface, and δ is known as the depth of penetration or the skin depth.

The phenomenon of skin effect gives a second reason for using laminated magnetic circuits. If the thickness of a plate is much more than twice the depth of penetration δ, the central region will carry very little flux. The material will be fully utilised if it is divided into laminations less than δ in thickness, for the flux density will then be fairly uniform across the lamination. The depth of penetration in silicon steel is about 1 mm at a frequency of 50 Hz, so the typical lamination thickness of 0.5 mm ensures that skin effect will not be significant.

1.6 The magnetic circuit

In the study of electromagnetic devices, it is often necessary to determine the magnetic field from a knowledge of the structure of the device and the magnitudes of the currents flowing in coils or other conductors. An accurate determi-

GENERAL PRINCIPLES 29

nation involves the solution of the partial differential equations of the electromagnetic field, a problem which is made more difficult by the non-linear properties of magnetic materials. Modern computational methods [6] can determine the field to any desired degree of accuracy, but a simple method of estimating the field is also required. The magnetic circuit concept provides an approximate method of solution which is good enough for many purposes and gives some immediate physical insight into the behaviour of magnetic structures.

The magnetic circuit concept

Figure 1.25 shows a closed iron core magnetised by a coil carrying a current. If the relative permeability of the iron is high, most of the magnetic flux will be

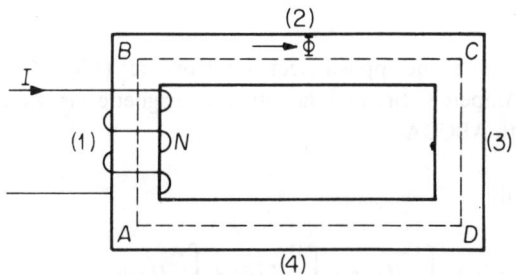

Figure 1.25 Simple magnetic circuit

confined to the iron. The flux Φ through any cross-section of the core will then be substantially the same; this follows from the fact that the flux of B out of any closed surface is zero; if there is no flux out of the sides of the iron, the flux entering a section must be equal to the flux leaving it. We thus have an analogy with a simple electric circuit (figure 1.26), in which the current i (which is the flux of the current density J) is the same for all cross-sections of the conductor. Corresponding to Kirchhoff's current law $\Sigma i = 0$ we have the flux law $\Sigma \Phi = 0$, and the structure of figure 1.25 is known as a *magnetic circuit*. We shall see that there is a very close analogy between the analysis of the electric and magnetic circuits.

Apply Kirchhoff's voltage law to the electric circuit of figure 1.26, following the path PQRST

$$e = -\Sigma v$$
$$= -(v_{PQ} + v_{QR} + v_{RS} + v_{ST}) \quad (1.69)$$
$$= \int_P^Q \boldsymbol{E} \cdot d\boldsymbol{s} + \int_Q^R \boldsymbol{E} \cdot d\boldsymbol{s} + \int_R^S \boldsymbol{E} \cdot d\boldsymbol{s} + \int_S^T \boldsymbol{E} \cdot d\boldsymbol{s} \quad (1.70)$$

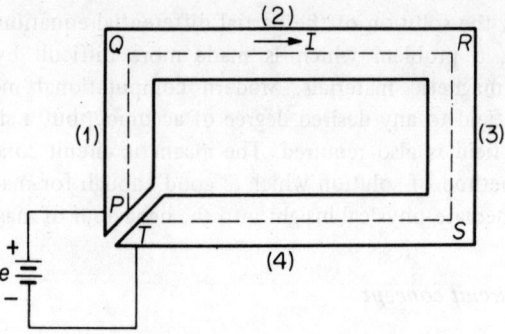

Figure 1.26 Simple electric circuit

The battery voltage e is the applied electromotive force (EMF) in the circuit.

Now apply Ampère's circuital law to the magnetic circuit of figure 1.25, following the path ABCDA

$$NI = \oint \boldsymbol{H} \cdot \mathrm{d}\boldsymbol{s}$$
$$= \int_A^B \boldsymbol{H} \cdot \mathrm{d}\boldsymbol{s} + \int_B^C \boldsymbol{H} \cdot \mathrm{d}\boldsymbol{s} + \int_C^D \boldsymbol{H} \cdot \mathrm{d}\boldsymbol{s} + \int_D^A \boldsymbol{H} \cdot \mathrm{d}\boldsymbol{s} \qquad (1.71)$$

Note the similarity between eqns (1.70) and (1.71). Just as the quantity

$$v_{PQ} = -\int_P^Q \boldsymbol{E} \cdot \mathrm{d}\boldsymbol{s}$$

is the electric potential difference between P and Q, the quantity

$$U_{AB} = -\int_A^B \boldsymbol{H} \cdot \mathrm{d}\boldsymbol{s}$$

is the magnetic potential difference between A and B. If we now put $NI = F$ (note that the symbol F in this context does not represent mechanical force), eqn (1.71) becomes

$$F = \oint \boldsymbol{H} \cdot \mathrm{d}\boldsymbol{s}$$
$$= -(U_{AB} + U_{BC} + U_{CD} + U_{DA})$$
$$= -\Sigma U \qquad (1.72)$$

and eqn (1.72) is the magnetic counterpart of Kirchhoff's voltage law (eqn 1.69). By analogy with the electromotive force, the quantity $F = NI$ is known as the magnetomotive force (MMF), measured in amperes (A). It follows that the units of magnetic potential, U, are also amperes.

Electric and magnetic circuit analogies

The analogy can be taken a stage further. In figure 1.26, suppose that limb (1) has a length l_1, a cross-sectional area A_1, and a constant conductivity σ_1. Since the electric field is practically uniform

$$\int_P^Q E \cdot ds = E_1 l_1$$

$$= \frac{J_1 l_1}{\sigma_1}$$

$$= \frac{i l_1}{\sigma_1 A_1}$$

$$= iR_1$$

where R_1 is the resistance of limb (1) given by

$$R_1 = \frac{l_1}{\sigma_1 A_1} \quad \text{ohms} \tag{1.73}$$

Thus eqn (1.70) becomes

$$e = i(R_1 + R_2 + R_3 + R_4) \tag{1.74}$$

In figure 1.25, suppose likewise that limb (1) has a length l_1, a cross-sectional area A_1, and a constant permeability μ_1 (where $\mu = \mu_0 \mu_r$). Since the magnetic field is practically uniform

$$\int_A^B H \cdot ds = H_1 l_1$$

$$= \frac{B_1 l_1}{\mu_1}$$

$$= \frac{\Phi l_1}{\mu_1 A_1}$$

$$= \Phi S_1$$

where S_1 is known as the *reluctance* of limb (1) and is given by

$$S_1 = \frac{l_1}{\mu_1 A_1} \quad \text{amperes/weber} \tag{1.75}$$

Thus eqn (1.71) becomes

$$NI = F = \Phi(S_1 + S_2 + S_3 + S_4) \tag{1.76}$$

The similarity between eqns (1.73) and (1.75), and between eqns (1.74) and (1.76) should be noted. It follows that the electric and magnetic circuits may be represented by the circuit diagrams of figures 1.27 and 1.28. Table 1.1 summarises the analogy between electric and magnetic circuits.

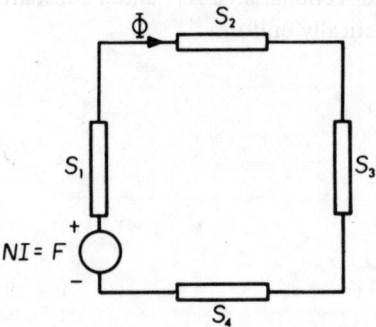

Figure 1.27 Magnetic circuit diagram

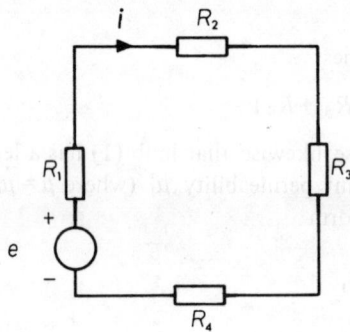

Figure 1.28 Electric circuit diagram

Simple magnetic circuits with airgaps

An example will illustrate the methods of analysing a simple magnetic circuit. Figure 1.29 shows an electromagnet consisting of an iron core with an airgap. The coil carries a current of 1 A, and we wish to find the number of turns required to set up a flux density of 1.2 T in the airgap. A simple method is to

GENERAL PRINCIPLES

Table 1.1. Electric and magnetic circuit analogies

Electric circuit	Magnetic circuit
Electromotive force e	Magnetomotive force F
Electric current i	Magnetic flux Φ
Electric potential difference v	Magnetic potential difference U
Resistance $R = v/i = l/\sigma A$	Reluctance $S = U/\Phi = l/\mu A$
Current law $\Sigma i = 0$	Flux law $\Sigma \Phi = 0$
Voltage law $\Sigma(e + v) = 0$	Circuital law $\Sigma(F + U) = 0$

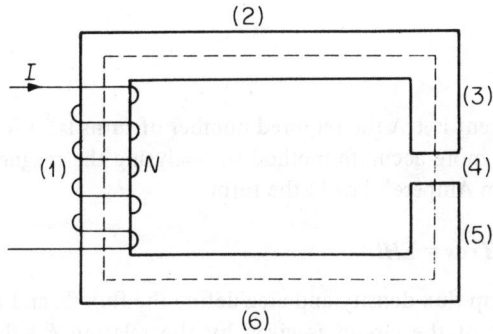

Figure 1.29 Electromagnet with an airgap

assume a constant value for the permeability of the core, and then to calculate the reluctances of the various parts of the magnetic circuit. If the core material is silicon steel, figure 1.21 shows that the relative permeability exceeds 3000 for a wide range of flux densities below the knee of the magnetisation curve. Table 1.2 gives the dimensions of the magnetic circuit and the calculated results with a value of 3000 for the relative permeability of the core. The flux in the core is given by

$$\Phi = BA = 1.2 \times 100 \times 10^{-6} \text{ Wb}$$
$$= 120 \text{ }\mu\text{Wb}$$

and the MMF is therefore

$$F = \Phi S = 120 \times 10^{-6} \times 17.9 \times 10^{6} \text{ A}$$
$$= 2170 \text{ A}$$

Table 1.2. Magnetic circuit calculation using constant permeability

Section of circuit	(1)	(2)	(3)	(4)	(5)	(6)
Material	iron	iron	iron	air	iron	iron
Relative permeability μ_r	3000	3000	3000	1	3000	3000
length l (mm)	150	250	74	2	74	250
Area A (mm^2)	80	125	100	100	100	125
Reluctance $S = l/\mu_0\mu_r A$ (MA/Wb)	0.05	0.53	0.20	15.9	0.20	0.53

$$\text{Reluctance of iron path} = 1.96 \text{ MA/Wb}$$
$$\text{Reluctance of airgap} = 15.9 \text{ MA/Wb}$$
$$\text{Total reluctance} = 17.9 \text{ MA/Wb}$$

Since the coil current is 1 A the required number of turns is 2170.

A second and more accurate method of analysing the magnetic circuit is to work directly from Ampère's law in the form

$$F = NI = \oint \boldsymbol{H} \cdot d\boldsymbol{s} = \Sigma Hl \qquad (1.77)$$

The specified airgap flux density and area define the flux Φ, and the flux density in any other part of the circuit is given by the relation $B = \Phi/A$. The corresponding value of H may be found from the magnetisation curve for the material used in that part of the circuit, and the sum of the Hl terms may then be computed. Table 1.3 shows the calculation; the value of H in the airgap is obtained from the relation $B = \mu_0 H$; the magnetisation curve of figure 1.20 is used for the iron parts; and the flux is given by $\Phi = BA = 120 \, \mu\text{Wb}$. The number of turns required on the coil is therefore 2250, and the total reluctance given by this calculation is

$$S = F/\Phi = 18.7 \text{ MA/Wb}$$

This reluctance is larger than the previous result of 17.9 MA/Wb. Table 1.3 shows that section (1) of the magnetic circuit has been driven into saturation, with a flux density of 1.5 T; this results in an excessive potential drop in the section, with a corresponding increase in the total reluctance. If the airgap flux density is reduced to 1.0 T, section (1) will come out of saturation with a flux density of 1.25 T. The whole iron path is then unsaturated, and a similar calculation gives a value of 16.7 MA/Wb for the total reluctance. This is smaller than the value of 17.9 MA/Wb calculated on the assumption of a constant permeability of 3000, because the relative permeability of most parts of the circuit is now greater than 3000, but the error is not large.

Table 1.3. Magnetic circuit calculation using the magnetisation curve

Section of circuit	(1)	(2)	(3)	(4)	(5)	(6)
Material	iron	iron	iron	air	iron	iron
Area A (mm^2)	80	125	100	100	100	125
Flux density $B = \Phi/A$ (T)	1.5	0.96	1.2	1.2	1.2	0.96
Magnetising force H (A/m)	1800	90	170	954×10^3	170	90
Length l (mm)	150	250	74	2	74	250
Potential drop Hl (A)	270	22.5	12.6	1910	12.6	22.5

Potential drop in iron = 340 A
Potential drop in airgap = 1910 A
Total potential drop = F = 2250 A

Linearity

In the example just considered the iron path is nearly 400 times the length of the airgap, but it contributes only about 10 per cent of the total reluctance of the magnetic circuit. Thus when the iron is unsaturated quite a small airgap will have a dominant effect, and any changes in the magnetic condition of the core will cause very little change in the total reluctance. Since $NI = \Phi S$, it follows that the flux will be proportional to the current if the total reluctance is constant, and this is the basis of the assumption of linearity made earlier in the chapter. There are two situations in which the relationship between flux and current is not even approximately linear. If there are no airgaps in the magnetic circuit, there will be no constant reluctance term to swamp the variable reluctance of the iron, and the relationship between flux and current will be determined by the magnetisation curve of the material. If there are airgaps but the iron is driven into saturation, the relative permeability of the iron will be low (see figure 1.21) and its reluctance may be comparable to the reluctance of the airgaps; the non-linear iron characteristic will again make its presence felt. The reader is invited to explore these effects in problem 1.8 at the end of the chapter.

Fringing and leakage

In the analysis of a magnetic circuit with an airgap, two assumptions have been made in order to simplify the calculation

(1) flux passes straight across the airgap, without spreading into the surrounding air
(2) there is no leakage of flux from the iron path into the surrounding air.

In practice, there is some spreading of the airgap flux (known as *fringing*), and leakage cannot be neglected when the airgap is large (that is, when its length is not negligible in comparison with the other air spaces between the iron parts). Both of these effects are illustrated in figure 1.30. It is possible to introduce fringing and leakage coefficients to take account of these effects, thereby extending the magnetic circuit theory to handle quite complex problems without having to resort to a full field analysis [10].

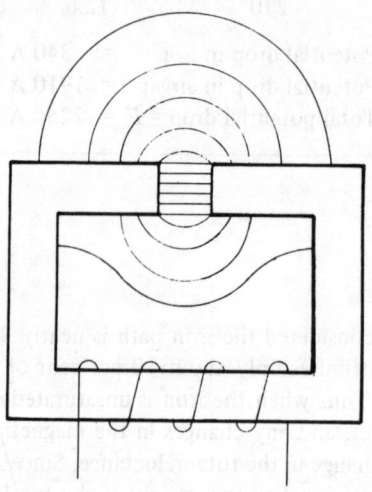

Figure 1.30 Fringing and leakage flux

Magnetic circuits with parallel paths

Figure 1.31 shows a magnetic circuit with parallel paths. A simple analysis is possible if we assume that the permeability is constant, and also neglect leakage and fringing. The circuit can then be represented by the diagram of figure 1.32, where S_a and S_i are the constant reluctances of the air and iron paths. Applying the flux law to the junction P gives $\Phi_3 = \Phi_1 + \Phi_2$, and applying the circuital law in the form $F = -\Sigma U = \Sigma \Phi S$ to meshes (1) and (2) gives

$$F = S_{i_3}(\Phi_1 + \Phi_2) + (S_{a_1} + S_{i_1})\Phi_1 \tag{1.78}$$

$$F = S_{i_3}(\Phi_1 + \Phi_2) + (S_{a_2} + S_{i_2})\Phi_2 \tag{1.79}$$

Equations (1.78) and (1.79) can be solved to find Φ_1 and Φ_2. Note that this is exactly analogous to applying Kirchhoff's voltage law in the form $e = \Sigma iR$ to the two meshes of a similar electric circuit. More complex magnetic circuits may be handled in a similar way, making full use of any relevant electric circuit theorems.

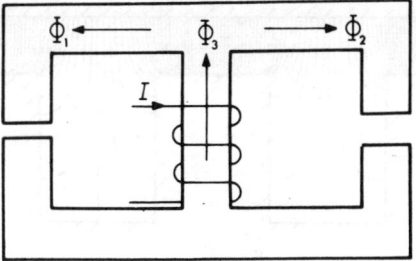

Figure 1.31 Magnetic circuit with parallel paths

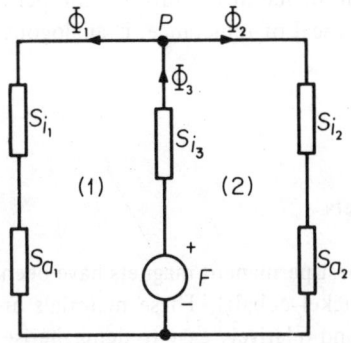

Figure 1.32 Magnetic circuit diagram for figure 1.31

Magnetic circuits: concluding remarks

The circuit analogy is a useful device for deducing the properties of magnetic structures from the more familiar properties of electric circuits. An EMF drives a current through an electric circuit against the resistance; an MMF drives a flux through a magnetic circuit against the reluctance, and the greater the reluctance the greater the MMF required to establish the flux. Just as current takes the path of least resistance, flux takes the path of least reluctance. In magnetic circuits with small airgaps the flux will be concentrated in the low-reluctance region of the gap, and the fringing field beyond the gap will fall away rapidly as the length of the air path increases. Figure 1.33 shows a type of structure commonly found in electrical machines, and it is immediately evident that the flux will be concentrated in the teeth, leaving a relatively weak field in the slots. An approximate calculation follows from the electric circuit rule for current division in parallel

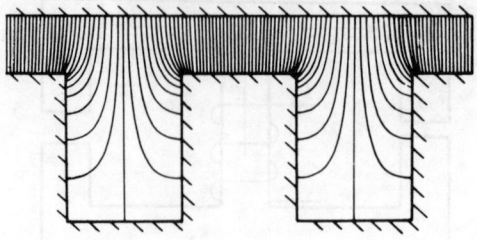

Figure 1.33 Field in electrical machine slots

branches: the flux will divide in the ratio of the permeances, where permeance (symbol Λ), the reciprocal of reluctance, is analogous to the conductance of an electric circuit.

1.7 Permanent magnets

For many years the best permanent magnets have been made from alloys such as Alnico (aluminium-nickel-cobalt). These materials are expensive, brittle, difficult to manufacture and relatively easy to demagnetise; they are not widely used in electrical machines. The development of ferrite permanent magnets has changed this position [11]; the material is much cheaper, it is easier to manufacture in special shapes, and it is much more difficult to demagnetise. Recently, new permanent-magnet materials based on alloys of cobalt and rare-earth elements such as samarium have been developed; they have much better magnetic properties than ferrites, but are very expensive. In spite of the cost they have been used in small electrical machines in increasing quantities, to the extent that supplies of the raw materials are becoming scarce. An exciting new material is neodymium-iron-boron [12], which has better magnetic properties even than samarium-cobalt, is considerably cheaper and has more abundant sources of raw materials. This material is likely to have a profound effect on the design of small and medium-sized electrical machines; its only major disadvantage at present is the loss of magnetisation with increasing temperature, which limits its use to below 125°C.

The magnetic circuit concept applies equally well when the source of the magnetic field is a permanent magnet instead of a current-carrying coil. Figure 1.34 shows such a circuit, where iron pole-pieces guide the flux from the magnet to the airgap. The permanent magnet has a length l_m and a cross-sectional area A_m; the airgap has a length l_g and a cross-sectional area A_g. To simplify the analysis, assume that the pole-pieces have negligible reluctance, and that there is

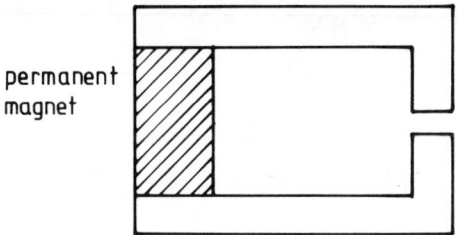

Figure 1.34 Permanent magnet with an airgap

negligible leakage or fringing. Since the flux is constant round the circuit, we have

$$B_m A_m = B_g A_g \tag{1.80}$$

There is no current, so Ampère's circuital law gives

$$0 = \oint H \cdot ds = H_m l_m + H_g l_g$$
$$= H_m l_m + B_g l_g / \mu_0 \tag{1.81}$$

Combining eqns (1.80) and (1.81) gives

$$B_m = -\mu_0 \frac{l_m A_g}{l_g A_m} H_m \tag{1.82}$$

This is a straight-line relationship between B_m and H_m imposed by the magnetic circuit. But the material of the permanent magnet also has its own inherent relationship between B_m and H_m, shown in the hysteresis loop for the material, and the portion of interest is the second quadrant where B is positive and H is negative. This part of the loop is termed the *demagnetisation characteristic*.

Figure 1.35 shows the demagnetisation characteristic for one type of neodymium-iron-boron permanent-magnet material [12]. The quantity B_r is known as the *remanence*; it is the flux density which remains in the material when the positive magnetising force H is reduced to zero. The quantity H_c is termed the *coercivity*; it is the negative magnetising force which must be applied to reduce the flux density to zero. Also shown in figure 1.35 is a straight line OM plotted from eqn (1.82) for a particular magnetic circuit. This line cuts the demagnetisation characteristic at P, which is the working point for the material in this magnetic circuit.

Suppose that the working point in figure 1.35 is moved from P to Q. This may be achieved by applying a reverse MMF with a current-carrying coil; or by introducing a large airgap into the magnetic circuit, with a characteristic line ON.

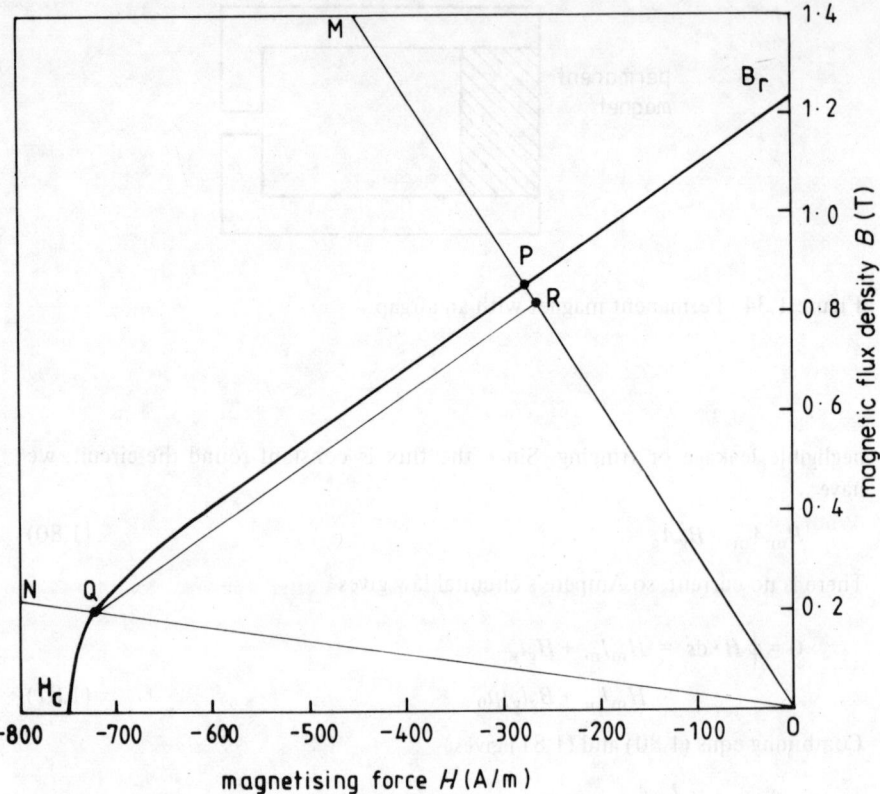

Figure 1.35 Demagnetisation characteristic for NeIGT 35 neodymium-iron-boron (I G Technologies, Inc.)

On removal of the demagnetising influence – the reverse MMF or the large airgap – the working point does not return along the original curve to P. Instead the working point traverses a line QR, known as a *recoil line*, to a new working point R on the original line OM. If a large demagnetising influence is applied, as can happen when an electrical machine is short-circuited, it is possible for the new working point R to be well below the original point P; this represents a partial demagnetisation of the material. With samarium–cobalt and some ferrites, the recoil line is almost indistinguishable from the original characteristic, and R is very close to P. With Alnico-type magnets, recoil is much more pronounced and severe demagnetisation can easily occur.

Modern high-performance permanent-magnet materials are expensive, and it is desirable to minimise the volume of the magnet given by

$$V_m = A_m l_m \tag{1.83}$$

Substituting for A_m and l_m from eqns (1.80) and (1.81) gives

$$V_m = - \frac{B_g^2 A_g l_g}{\mu_0 B_m H_m} \tag{1.84}$$

Thus for a given airgap flux density B_g and volume $A_g l_g$, the volume of the magnetic material will be a minimum when the energy product $B_m H_m$ is a maximum. The value of B_m which gives the maximum energy product can be found from a graph of $B_m H_m$ against B_m; this defines the optimum working point, and the magnet dimensions can be calculated from eqns (1.80) and (1.81). Usually the desired airgap flux density B_g is greater than the optimum magnet flux density B_m; consequently the area A_g must be less than A_m, and the pole-pieces act as flux concentrators.

Problems

1.1. A flat copper plate is held in a vertical plane and then allowed to fall through the gap between the poles of a magnet. The plate is much wider than the poles, and any induced currents which flow in the part of the plate between the poles are assumed to find return paths of negligible resistance in the rest of the plate. The magnet poles are square, and the magnetic field may be assumed to be uniform and confined to the area of the poles. Show that the plate will experience a retarding force proportional to its velocity, and calculate the magnitude of the force when the velocity is 10 m/s. The thickness of the plate is 5 mm; the resistivity of copper is 1.68×10^{-8} Ω m; the magnet poles are 100 mm square; and the magnetic flux density is 1.0 T.

1.2. In the Faraday disc machine (section 1.2) there will be a torque on the disc when current flows between the centre and the periphery. By considering the torque on an elementary annular ring of the disc, show that the total torque is given by the expression

$$T = \tfrac{1}{2} B i a^2 \text{ newton metres}$$

where B is the magnetic flux density in teslas, i is the current in amperes and a is the radius of the disc in metres. Hence show that the mechanical power output from the disc is equal to the electrical power input when the machine runs as a motor. The resistance of the disc may be ignored.

1.3. Figure 1.36 shows an extract from a patent specification [13] for an improved type of DC generator. The rotor R is magnetised by the field coil F, and thus forms a rotating magnet. The inventors claim that the rotating field of this magnet will induce an EMF in a stationary conductor such as C, which forms one side of a rectangular coil. Several coils may be connected in series to increase the output voltage of the machine.

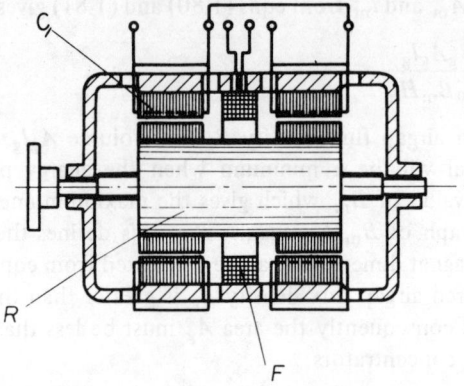

Figure 1.36 Proposed DC generator (reproduced with the permission of the Controller of Her Majesty's Stationery Office)

Some engineers have expressed doubts as to whether the machine will actually work. Settle the question by the application of Faraday's law.

1.4. Show that there can be no magnetic field outside a coaxial cable carrying a current. Hence show that an alternating magnetic field due to currents in other conductors will induce no voltage in the cable circuit.

1.5. A steel ring is uniformly wound with a coil of 1000 turns, and a flux density of 1.5 T is produced in the ring by a coil current of 3 A. If the ring has a mean diameter of 0.2 m and a cross-sectional area of 0.001 m^2, calculate the inductance of the coil.

The ring is divided into two equal parts by making cuts in the iron. The width of each cut is 2 mm, and the current in the coil is increased to maintain the original flux density of 1.5 T. Calculate (a) the new value of the coil current; (b) the new coil inductance; (c) the force of attraction between the two halves of the ring.

1.6. The device shown in figure 1.17 will function as an elementary form of AC motor when the coil carries a current of the form $i = I_m \cos \omega t$. The inductance of the coil depends on the rotor position and may be assumed to follow the law $L = L_1 + L_2 \cos 2\theta$, where θ is the angular position of the rotor. If the rotor revolves with a steady angular velocity ω_r, so that $\theta = \omega_r t + \phi$, show that the torque on the rotor is given by the expression

$$T = -\tfrac{1}{2} L_2 I_m^2 (1 + \cos 2\omega t) \sin (2\omega_r t + 2\phi)$$

Show that the torque will have an average value of zero unless $\omega_r = \omega$, and obtain an expression for the average value when this condition holds.

1.7. The following figures were obtained for the power loss in the core of a transformer at different frequencies, with the maximum value of the flux density held constant

frequency (Hz)	35	40	45	50	55	60	65	70
power loss (W)	46	54	62	70	78	87	96	105

If the loss is made up of hysteresis and eddy current components, show that the total loss should be related to the frequency by an expression of the form

$$P = Af + Bf^2$$

where A and B are constants. By plotting a graph of P/f against frequency, determine the constants A and B for the transformer, and hence find the values of the hysteresis and eddy current components of the core loss at a frequency of 50 Hz.

1.8. Assume that the ring in problem 1.5 is made from silicon steel, with the magnetisation curve shown in figure 1.20. Calculate the values of coil current required to give a number of values of flux density ranging from 0 to 1.5 T, (a) for a solid ring, (b) for a ring with a single airgap of 1 mm. Hence plot graphs of flux against current and inductance against current for the two cases.

1.9. For the magnetic circuit of figure 1.34, suppose that the demagnetisation characteristic of the permanent magnet is a straight line. Show that the magnet may be represented by an MMF $F_m = H_c l_m$ in series with a reluctance $S_m = l_m/\mu_m A_m$, where l_m is the length of the magnet, A_m is its cross-sectional area and $\mu_m = B_r/H_c$ is the slope of the demagnetisation characteristic.

References

1. P. Dunsheath, *A History of Electrical Engineering* (London: Faber, 1962).
2. A. D. Appleton, 'Superconducting DC machines', *IEEE Trans. Magnetics*, **MAG-11** (1975), pp. 633-9.
3. G. W. Carter, *The Electromagnetic Field in its Engineering Aspects*, 2nd ed. (London: Longman, 1967).
4. K. J. Binns, 'Flux cutting or flux linking', *J. IEE*, **9** (1970), p. 259.
5. J. A. Stratton, *Electromagnetic Theory* (New York: McGraw-Hill, 1941).
6. P. P. Silvester and R. L. Ferrari, *Finite Elements for Electrical Engineers* (Cambridge University Press, 1983).
7. C. J. Carpenter, 'Surface integral methods of calculating forces on magnetized iron parts', *Proc. IEE*, **107C** (1960), pp. 19-28.
8. F. Brailsford, *Physical Principles of Magnetism* (London: Van Nostrand, 1966).
9. R. L. Stoll, *The Analysis of Eddy Currents* (Oxford University Press, 1974).
10. Department of Electrical Engineering, Massachusetts Institute of Technology, *Magnetic Circuits and Transformers* (Cambridge, Mass: MIT Press, 1943).

11 M. McCaig, *Permanent Magnets in Theory and Practice* (Plymouth: Pentech Press, 1977).
12 I G Technologies, Inc., *NeIGT Permanent Magnet Material* (I G Technologies, Inc., 1984).
13 British Patent 917 263 (1963).

2 Direct Current Machines

2.1 Introduction

Historically, DC machines were the first to be developed because the only available electrical power source was the DC voltaic cell. The advantages of alternating current were later recognised, and the invention of the induction motor was an important factor in securing acceptance of the alternating current system. The two main types of AC machine (synchronous and induction machines) are structurally simpler than DC machines; but the theory of AC machines is inherently more complex than the theory of DC machines, and we therefore adopt the historical order in developing the principles.

The homopolar machine mentioned in section 1.2 is a DC machine; in fact it is a pure DC machine, for we shall see that the conventional machine generates an alternating voltage which is rectified mechanically by the commutator. Before the advent of cryogenics, the homopolar machine was suitable only for certain low-voltage heavy-current applications, and the development of DC machines followed a different path. If a coil rotates in a magnetic field (figure 2.1), the flux linking the coil will be an alternating quantity, and an alternating EMF will be induced in the coil. In figure 2.1 connection is made to the coil by brushes bearing on sliprings; an alternating voltage is developed at the terminals (figure 2.2) which reverses with every half revolution of the coil. Suppose that instead of being connected to separate rings, the two ends of the coil are connected to the two parts of a divided ring (figure 2.3); this reverses the connections to the

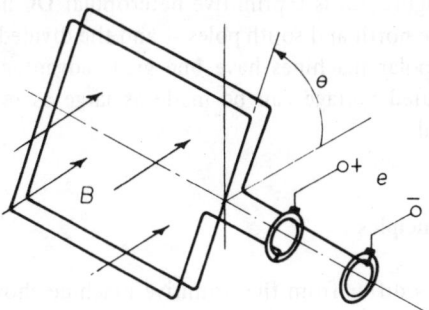

Figure 2.1 Elementary AC machine

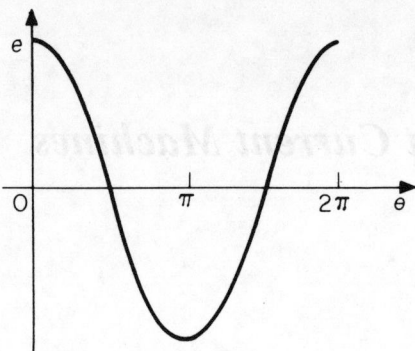

Figure 2.2 Generated EMF for the elementary AC machine

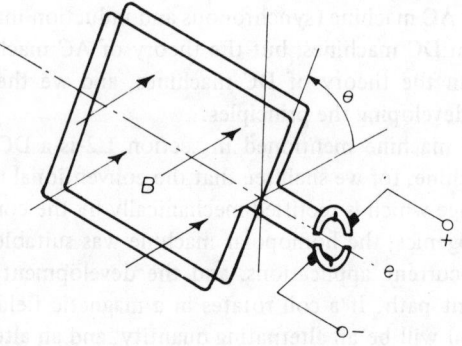

Figure 2.3 Elementary DC machine

coil every half revolution, so the terminal voltage is now unidirectional (figure 2.4). The device in figure 2.3 is a primitive heteropolar DC machine — the coils move under successive north and south poles — and the divided ring is a primitive commutator. Heteropolar machines have one great advantage over homopolar machines: the generated voltage can be made as large as required by winding more turns on the coil.

2.2 Fundamental principles

Practical DC machines differ from the primitive machine shown in figure 2.3 in one important respect: the active (armature) coils are wound on an iron cylinder, in order to reduce the length of the airgap in the magnetic circuit. The magnetic

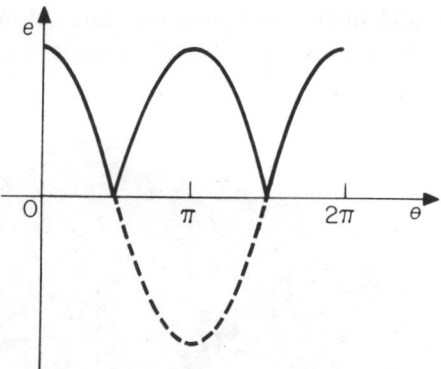

Figure 2.4 Generated EMF for the elementary DC machine

field is produced either by coils wound on iron poles or by permanent magnets; a small airgap will reduce the MMF — and hence the coil current or the magnet volume — required for a given flux density. Figure 2.5 shows the construction of a small wound-field machine; the armature coils are placed in slots in a cylindrical

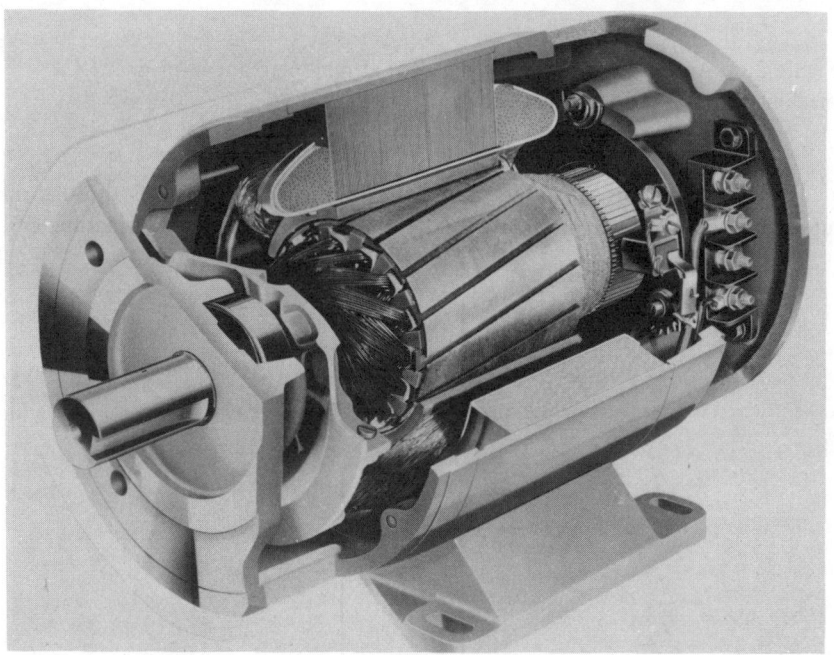

Figure 2.5 Construction of a small wound-field DC machine (GEC Small Machines Ltd)

iron core, and the faces of the field poles are curved to match the armature shape. Figure 2.6 shows the armature and field magnets for a small permanent-magnet machine.

Figure 2.6 Armature and field magnets for a small permanent-magnet DC machine (GEC Small Machines Ltd)

The simplest approach to the principles of the DC machine is through an idealised model, shown in figure 2.7, which has a single-turn armature coil

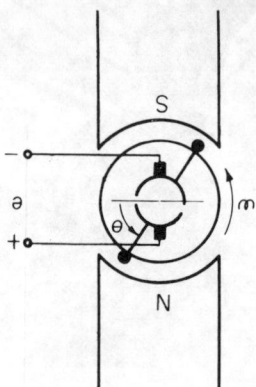

Figure 2.7 Simple DC machine

wound on the surface of a smooth iron cylinder. This model will be analysed to obtain the fundamental machine equations, and it will be shown that the equations also hold for the practical machine.

Voltage and torque equations for a simple model

Take the model shown in figure 2.7; assume that the magnetic field in the airgap is purely radial, and uniform along the length of the armature. The flux density B will, of course, vary with the angular position round the airgap, as shown in figure 2.8. Only the coil sides in the airgap will be considered; assume that the magnetic field outside the gap is negligible, so that the rest of the coil contributes nothing to the EMF or the torque.

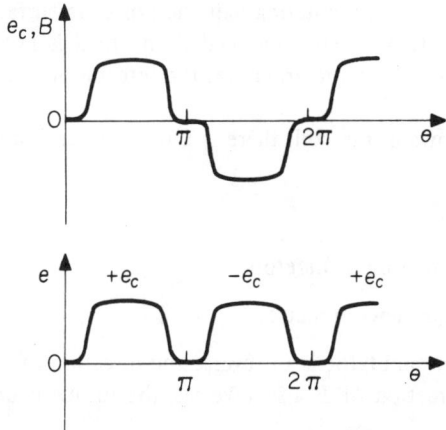

Figure 2.8 Flux density and generated EMF for the simple DC machine

As the armature rotates, the coil sides move in a magnetic field; an EMF will be generated in each conductor, given by eqn (1.5)

$$e = Blu \quad \text{volts}$$

The EMF developed in the whole coil is therefore

$$e_c = 2Blu = 2Blr\omega \quad \text{volts} \qquad (2.1)$$

where l is the length of the armature in metres, r is the radius in metres and ω is the angular velocity in radians/second. The induced EMF thus varies with the angular position of the coil in the same way as the flux density, as shown in figure 2.8. The action of the commutator is to invert the negative half cycles of the coil EMF, so the terminal voltage is given by

$$e = 2\omega lr |B(\theta)| \qquad (2.2)$$

as shown in figure 2.8. This is the instantaneous voltage, and the average value is

$$e_{av} = \frac{1}{\pi}\int_0^\pi e\, d\theta = \frac{1}{\pi}\int_0^\pi 2\omega lrB\, d\theta \qquad (2.3)$$

Now $lr\, d\theta = da$, an element of area of the armature surface. Equation (2.3) therefore becomes

$$e_{av} = \frac{2}{\pi}\omega \int_0^\pi Blr\, d\theta = \frac{2}{\pi}\omega \int_{\theta=0}^{\theta=\pi} B\, da$$

$$= \frac{2}{\pi}\Phi\omega \text{ volts} \qquad (2.4)$$

where Φ is the magnetic flux entering half the armature surface from one field pole. Note that eqn (2.4) is still obtained if the field is not purely radial or uniform along the length of the armature; the integration is simply more complicated.

If a current i_c flows in the coil, there is a force on each conductor given by eqn (1.14)

$$F = Bli_c \text{ newtons}$$

The torque on the armature is therefore

$$T = 2Fr = 2Bli_c r \text{ newton metres} \qquad (2.5)$$

The action of the commutator is to reverse the direction of i_c every half revolution; since the direction of B also reverses, the torque is unidirectional, and may be written as

$$T = 2ilr\,|B(\theta)| \qquad (2.6)$$

where i is the current in the armature circuit. The average torque is therefore

$$T_{av} = \frac{1}{\pi}\int_0^\pi 2ilB\, d\theta$$

$$= \frac{2}{\pi}\Phi i \text{ newton metres} \qquad (2.7)$$

Armature windings and commutator

The generated voltage of the simple DC machine is unidirectional, but far from constant. The performance can be improved by adding more conductors and commutator segments. Consider the effect of a second coil at right angles to the first (figure 2.9). The coil induced-EMF waveforms are shown in figure 2.10, together with the EMF e appearing at the terminals. There is an improvement in the output waveform, but each coil is now used for only half the time.

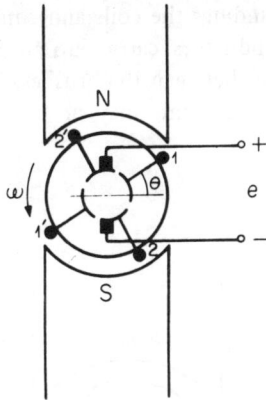

Figure 2.9 DC machine with two armature coils

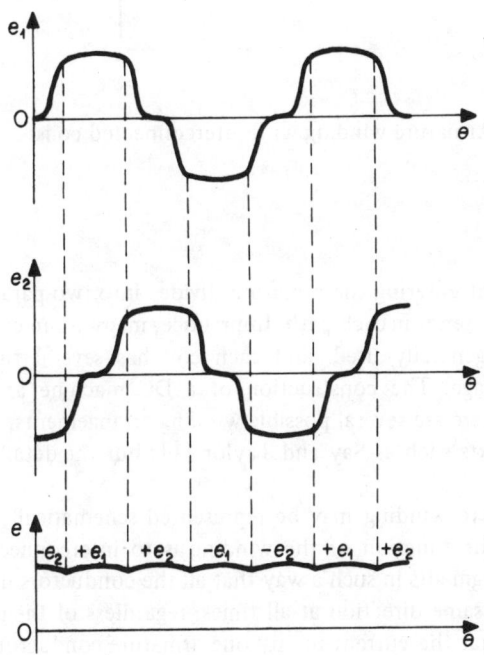

Figure 2.10 Generated EMF for the two-coil machine

In practical armature windings the coils and commutator segments are interconnected so that the conductors carry current all the time, and there are usually several coils in series between the brushes. Figure 2.11 shows how four coils and four commutator segments may be used to achieve this result; note

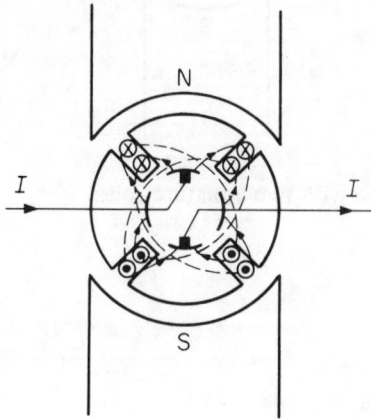

Figure 2.11 Armature winding with interconnected coils

that the current entering the armature divides into two parallel paths, and there are two coils in series in each path. In practice, many more coils and commutator segments are generally used, and each coil has several turns to increase the generated voltage. The construction of a DC machine armature is shown in figure 2.12. There are several possible winding arrangements, which are described in standard texts such as Say and Taylor [1]; but the details do not concern us here.

Any armature winding may be represented schematically by the diagram of figure 2.13. The function of the winding is to interconnect the coils and the commutator segments in such a way that all the conductors under one pole carry current in the same direction at all times, regardless of the motion of the armature. Notice that the current in any one armature conductor must reverse when the conductor passes the magnetic neutral axis between the poles. The currents in the armature coils are therefore alternating quantities, and the iron core of the armature is invariably laminated to reduce eddy current losses.

DIRECT CURRENT MACHINES 53

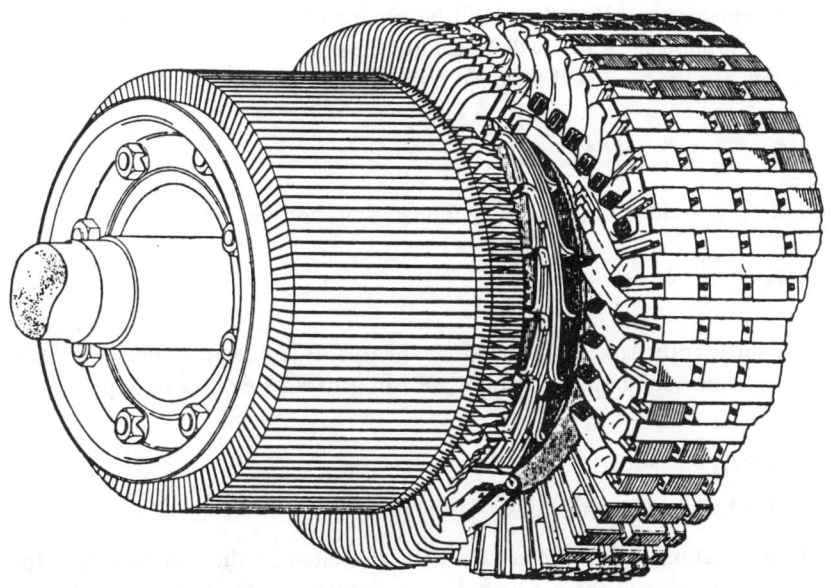

Figure 2.12 Construction of a DC machine armature (reproduced from *Electrical Machines* by A. Draper, Longman, 2nd edition, 1967)

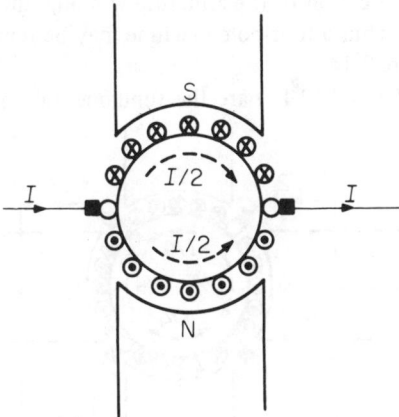

Figure 2.13 Schematic representation of a DC machine armature

General equations of the DC machine

In any armature winding there will be groups of coils in series between the brushes; the induced EMFs will be additive, and if there are n conductors in series between the brushes the average induced EMF will be

$$e_{av} = \frac{2n}{\pi} \Phi \omega \text{ volts} \tag{2.8}$$

Since these n conductors each carry the same current, the average torque will be

$$T_{av} = \frac{2n}{\pi} \Phi i \text{ newton metres} \tag{2.9}$$

With a sufficiently large number of conductors, the generated voltage and the torque will be very nearly constant; we can generalise eqns (2.8) and (2.9) to give

$$e_a = K_a \Phi \omega \text{ volts} \tag{2.10}$$

$$T = K_a \Phi i_a \text{ newton metres} \tag{2.11}$$

In these equations, e_a is the steady EMF generated by the armature; i_a is the armature current, and the constant K_a is a property of the particular armature winding. In the derivation of these equations we have not assumed that ω or i_a is a constant quantity, and they hold for transient as well as steady-state conditions.

More than two field poles may be employed, simply by repeating the N–S sequence as many times as desired round the periphery of the armature, with a corresponding modification of the armature winding; the fundamental principles remain unchanged. Thus a four-pole machine may be represented by the schematic diagram of figure 2.14.

Equations (2.10) and (2.11) are the fundamental equations of the machine,

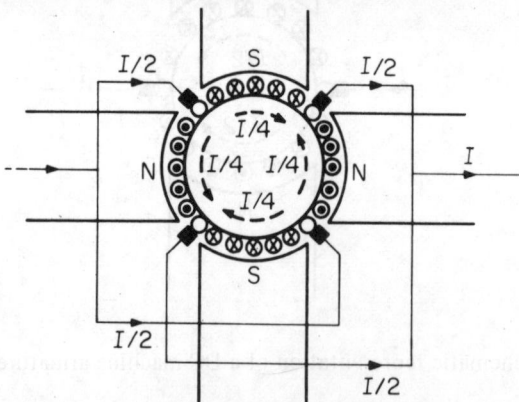

Figure 2.14 Four-pole DC machine

and we shall shortly deduce some of the interesting and useful characteristics of DC machines from them. Their simplicity is striking, and it conceals the inherent complexity of the commutation process. The action of a well-designed armature winding and commutator is to convert the alternating quantities in the armature coils into steady quantities at the brush terminals. The coils necessarily possess inductance, and the reversal of current in an inductive circuit is accompanied by an induced EMF or 'reactance voltage' which can cause sparking at the commutator. DC machines are therefore usually fitted with auxiliary poles (known as *interpoles*) to improve the commutation. These poles are placed midway between the main poles, and are wound with coils connected in series with the armature; their function is to induce an EMF which opposes the reactance voltage in the armature coils undergoing commutation. Interpoles do not affect the fundamental eqns (2.10) and (2.11), and they will not be considered further.

Field system and magnetisation curve

In permanent-magnet machines the field flux Φ is constant for normal operation, and there is nothing to add to eqns (2.10) and (2.11). But in wound-field machines it is necessary to consider how the pole flux is produced. Figure 2.15 is a schematic diagram of the structure of a wound-field DC machine without interpoles; the coils wound on the field poles are termed the *field* or *excitation winding* of

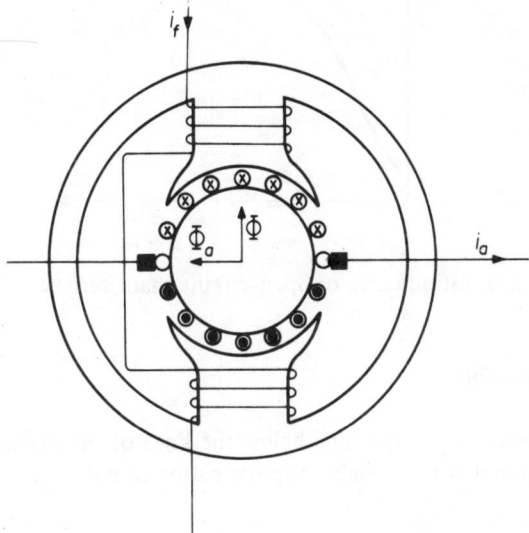

Figure 2.15   Wound-field DC machine

the machine. A current i_f flowing in the field winding will produce a pole flux Φ, as shown. With no armature current flowing, Φ will be a function of i_f only, and we may put $\Phi = \Phi(i_f)$; a graph of Φ against i_f (or $N_f i_f$, the field MMF) is known as the *magnetisation curve* of the machine. At constant speed, eqn (2.10) gives $e_a \propto \Phi$; a graph of e_a against i_f for zero armature current is known as the *open-circuit characteristic* of the machine, and this has the same shape as the magnetisation curve. Figure 2.16 shows a typical curve, and it will be seen that the middle portion is practically linear. In this region the iron is unsaturated, and the airgap reluctance is dominant in the magnetic circuit. The reluctance of the iron path increases rapidly with saturation, and this explains the shape of the curve for high values of field current. There is usually some remanent magnetisation of the iron, which accounts for the departure from linearity at low values of field current. Figure 2.16 is actually an over-simplification, for the magnetic circuit exhibits hysteresis; the curve for decreasing excitation will be slightly different from the curve for increasing excitation. The open-circuit characteristic may be regarded as the mean of the two curves.

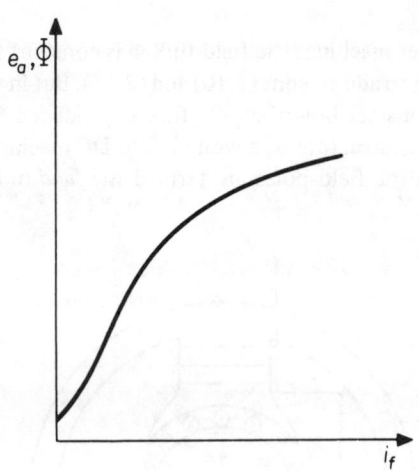

Figure 2.16 Magnetisation curve or open–circuit characteristic

Linear approximation

It is usual to operate the machine below the knee of the magnetisation curve, and for this region it is a reasonable approximation to put

$$\Phi = K_f i_f \tag{2.12}$$

The fundamental equations for this linear model then become

$$e_a = K i_f \omega \text{ volts} \tag{2.13}$$

$$T = K i_f i_a \text{ newton metres} \tag{2.14}$$

where $K = K_a K_f$. These are the equations normally used in the analysis of systems containing DC machines.

Armature reaction

A current i_a flowing in the armature will produce a flux Φ_a, at right angles to Φ; this is known as the *armature reaction flux*, and by itself it will produce no torque or EMF. If the magnetic circuit were linear, there would be no interaction between Φ and Φ_a; in practice, Φ_a may cause local saturation of the magnetic circuit, and this will reduce the value of Φ for a given field current i_f. Thus $\Phi = \Phi(i_f, i_a)$, and in some applications this non-linear dependence of Φ on i_f and i_a must be considered. It is often sufficient, however, to ignore the effect of armature reaction and to take eqns (2.13) and (2.14) as the basic machine equations. The armature reaction flux can adversely affect the commutation, especially when the armature current changes rapidly. To overcome this difficulty, DC machines are sometimes fitted with compensating windings. These take the form of conductors embedded in slots in the field pole faces; they are connected in series with the armature, but carry current in the opposite direction so as to cancel the armature reaction flux.

DC machine action in terms of magnetic forces

The concept of armature reaction suggests an alternative physical picture for the mechanism of torque production in the DC machine. The existence of an armature reaction flux implies magnetisation of the armature iron, which may be represented by N and S poles, and the resultant magnetic field produced by the armature and field poles is shown in figure 2.17. From the Maxwell stress concept (or the properties of magnetic poles) it follows that there will be a torque on the armature tending to rotate its poles into alignment with the field poles. The armature winding and commutator, however, ensure that the magnetic axis of the armature remains fixed in space while the armature material revolves; a steady torque is therefore developed, which is unaffected by the rotation of the armature.

Slotted armature

In practice, as will be seen from figure 2.12, the armature conductors are placed in slots in the armature core. This profoundly alters the electromagnetic action of the machine, for the magnetic field in a slot is relatively weak (see figure 1.33) and the force on a conductor is reduced in consequence. It may be shown [2, 3] that the reduction in the conductor force is exactly compensated by

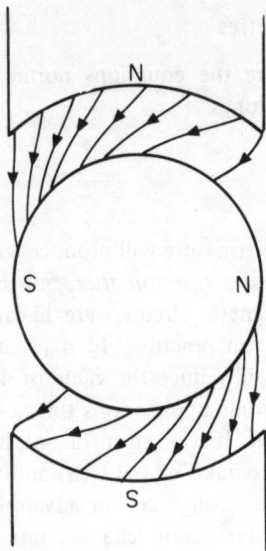

Figure 2.17 Magnetic field in a DC machine

forces acting on the slotted iron structure of the armature; the total torque is still given by eqn (2.11). A qualitative explanation is that the magnetic field pattern around the armature (figure 2.17) is, on average, the same whether or not the conductors are placed in slots, and the torque produced by the Maxwell stress will therefore be unchanged.

The EMF generated by the machine is still given by eqn (2.10), as may be seen by applying Faraday's law to an armature coil; the change of flux through the coil during one revolution of the armature is the same regardless of whether the coil sides are in slots or on the surface. The flux cutting rule $e = Blu$ will give different results if the field in the vicinity of the conductor is used for B; but the dangers inherent in the use of this rule have already been mentioned (see section 1.3).

An important principle is thus established: the basic machine equations are the same for a slotted armature as for a smooth cylindrical armature, provided that the total pole flux Φ (and hence the average value of B in the airgap) is unchanged. This result will be used in later chapters to simplify the analysis of AC machines.

2.3 Energy conversion and losses

Generators convert energy from mechanical to electrical form; motors perform the inverse operation, and the efficiency of energy conversion is often an important consideration. Since the efficiency is directly related to the energy loss in the machine, the sources of loss are matters of considerable importance to the machine designer. The machine user also needs to be aware of these losses, and it is useful to discuss them briefly before examining the other characteristics of DC machines.

Generators and motors: sign conventions

A DC machine with its armature connected to a steady voltage source V_a is shown symbolically in figure 2.18. The armature circuit (conductors, commutator and brushes) will have a resistance represented by R_a, and the field

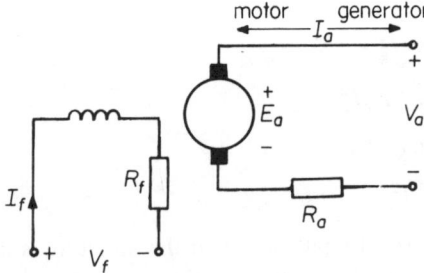

Figure 2.18 DC machine: circuit convention

winding will have a resistance R_f. If the steady generated voltage E_a is less than V_a, the source will supply power to the machine, which therefore acts as a motor. If we choose the direction of current flow so that the armature current I_a is positive under these conditions, then

$$V_a = E_a + R_a I_a \qquad (2.15)$$

If $E_a > V_a$, the source will absorb power from the machine, which therefore acts as a generator. With the opposite direction for positive I_a, we now have

$$E_a = V_a + R_a I_a \qquad (2.16)$$

Generator and motor action differ only in the directions of current and torque, and we adopt the convention that both the torque and the armature current will be positive when the machine is operating as a motor. The correct armature voltage equation is therefore eqn (2.15), not eqn (2.16); negative values of I_a and T will indicate that the machine is operating as a generator.

Losses and efficiency

The losses in a DC machine are essentially the same whether the machine operates as a generator or a motor, and motoring operation will be assumed for the rest of this section. Consider a DC machine with its armature connected to a voltage source V_a as shown in figure 2.18. With steady-state conditions the basic machine equations are

$$E_a = K_a \Phi \omega \qquad [2.10]$$

$$T = K_a \Phi I_a \qquad [2.11]$$

and we also have the motor armature equation

$$V_a = E_a + R_a I_a \qquad [2.15]$$

Multiplication of eqn (2.11) by ω, and eqns (2.10) and (2.15) by I_a, gives

$$\omega T = K_a \Phi I_a \omega \qquad (2.17)$$

$$\begin{aligned} V_a I_a &= E_a I_a + R_a I_a^2 \\ &= K_a \Phi \omega I_a + R_a I_a^2 \end{aligned} \qquad (2.18)$$

From eqns (2.17) and (2.18)

$$V_a I_a = \omega T + R_a I_a^2 \qquad (2.19)$$

showing that the electrical input power to the armature is divided between the gross mechanical power ωT and the resistive loss $R_a I_a^2$. The mechanical output from the motor shaft will be less than ωT, because some of the torque T (known as the *gross*, or *electromagnetic*, *torque*) will be absorbed in rotational losses. The flow of energy through the machine may be traced as follows

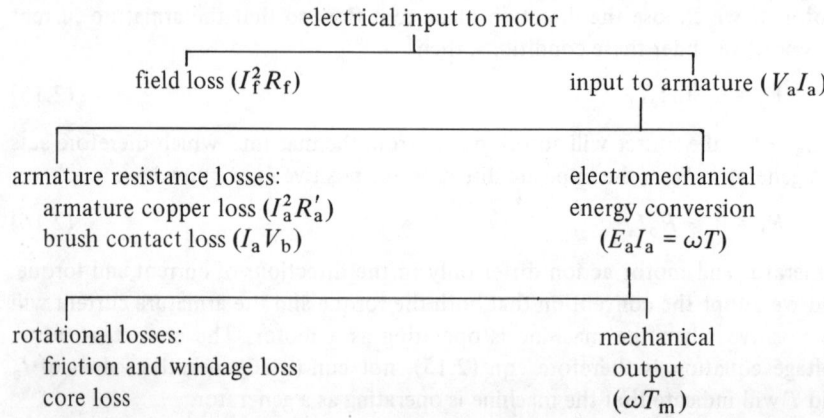

Components of loss

The losses in the machine comprise the field and armature resistance losses and the rotational losses. Although the armature resistance loss is usually represented by the expression $R_a^2 I_a$, it should strictly be separated into the two components shown above because the brush contact resistance is non-linear. The voltage drop between the brushes and the commutator segments has an approximately constant value V_b over a wide range of currents, with a typical value of 2 volts for a pair of normal carbon brushes.

The two components of rotational loss are quite different in character. 'Friction and windage' accounts for all the mechanical and aerodynamic losses associated with the rotation of the armature, and varies roughly as the square of the speed. 'Core loss' includes the eddy-current and hysteresis loss in the laminated iron armature core, together with a similar surface loss in the field poles. The latter is a consequence of placing the armature conductors in slots; as the slots move past the field poles, the resulting local variations in the magnetic flux density produce eddy-current and hysteresis losses on the pole faces. A laminated construction is often used to minimise the eddy-current component of pole-face loss. The total core loss varies in a complex way with the speed, armature current and field flux.

Efficiency

An important property of an electrical machine is its efficiency under specified operating conditions. The efficiency η is defined in the usual way as

$$\eta = \frac{\text{output power}}{\text{input power}} \tag{2.20}$$

$$= \frac{\text{input power} - \text{losses}}{\text{input power}}$$

$$= 1 - \frac{\text{losses}}{\text{input power}} \tag{2.21}$$

Because of the difficulty of measuring input and output power accurately, the direct evaluation of efficiency implied by eqn (2.20) is seldom used; instead the losses are determined (usually from a number of different tests) and the efficiency is calculated from eqn (2.21).

2.4 DC generators

As sources of DC power, DC generators have been largely replaced by controlled semiconductor rectifiers, and a detailed treatment of their characteristics would be out of place in this book. Three types of generator, however, are worthy of mention: the permanent-magnet generator; the separately excited generator, in which the field winding is supplied from a separate power source; and the shunt generator or dynamo, which supplies its own excitation power.

Permanent-magnet generator

In a permanent-magnet DC machine the field flux Φ is substantially constant, and eqn (2.10) shows that the generated voltage e_a is directly proportional to the speed ω. Small machines of this kind, known as *tachogenerators*, are built specifically for speed measurement, and are widely used in feedback control systems. They require careful design to ensure that the field flux does not change with temperature or time.

Separately excited generator

When a wound-field DC machine is used as a source of power it is driven at a constant speed ω, and the field winding is connected to a voltage source V_f (figure 2.18). There will be a constant flux Φ, and the armature generated voltage has a constant value E_a given by eqn (2.10). With no load connected to the armature terminals, $V_a = E_a$; thus

$$V_a = K_a \omega \Phi \tag{2.22}$$

The relationship between the open-circuit terminal voltage V_a and the field current I_f is given by the open-circuit characteristic (figure 2.16). In the linear region of the characteristic we may put

$$V_a = E_a \approx K\omega I_f = \frac{K\omega}{R_f} V_f \tag{2.23}$$

showing that the output voltage is approximately proportional to the input voltage or current. The excitation power $V_f I_f$ is only a few per cent of the output power $V_a I_a$ when the machine is connected to a load; a DC generator may be regarded as a power amplifier, and has been widely used as such in control schemes such as the Ward–Leonard system [1].

Shunt generator

Instead of being connected to a separate voltage source, the field winding of a shunt generator is connected in parallel with the armature, so that the armature itself supplies the excitation current (figure 2.19). Frequently there is a variable resistor in series with the field, and total resistance will be denoted by R_f.

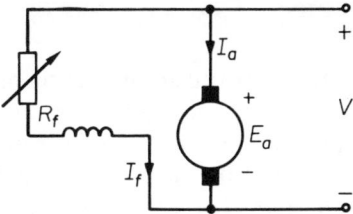

Figure 2.19 Shunt generator or motor

The operation of a shunt generator may be understood by drawing on the open-circuit characteristic of the machine a straight line representing the voltage/current characteristics of the resistance R_f (figure 2.20). The point at which this line intersects the open-circuit curve gives the no-load terminal voltage of the

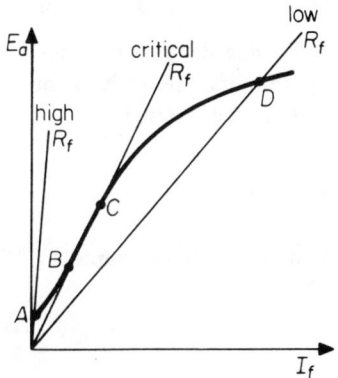

Figure 2.20 Operation of the shunt generator

machine, for the current flowing in the resistance is then just equal to the field current required to maintain that value of terminal voltage. When the resistance is high, the line intersects the curve at a point such as A, and the armature voltage is very small. As the value of R_f is reduced, the point of intersection moves up the curve until a critical value of R_f is reached; for lower values of R_f the operating point is at a position such as D in the saturation region. When R_f is equal to the critical value the generator is unstable, and very small changes in conditions can move the operating point from B to C. Shunt generators are normally designed to work well into the saturation region, so that the operating point is stable and the output voltage substantially constant.

2.5 DC motors

The great virtue of the DC motor is that three useful operating characteristics may be obtained by using the field winding in different ways. In the separately excited motor the armature and field windings are supplied from independent voltage sources, and the speed of the motor may be controlled by varying the voltage applied to the armature. The shunt motor has its field connected in parallel with the armature; this gives an almost constant speed, independent of voltage, over a wide range of loads. The series motor has its field connected in series with the armature, and this gives a characteristic in which the speed is inversely related to the torque load on the motor.

The essential features of DC motor performance may be deduced fairly readily if the following idealising assumptions are made

(a) the armature resistance R_a is neglected
(b) rotational losses are neglected
(c) magnetic non-linearity is ignored
(d) the motor operates under steady-state conditions.

As will be seen in the following discussion, these assumptions are not always appropriate; they greatly simplify the analysis, but their use must always be recognised and the implications appreciated.

Ideal motor characteristics

With the motor connected to a voltage source V_a (figure 2.18) the armature circuit equation is

$$V_a = E_a + R_a I_a \qquad [2.15]$$

The power loss in the armature resistance R_a is generally small in comparison with the input power to the motor, at least for machines with power ratings above 1 kW. It follows that the voltage drop $I_a R_a$ is usually small in comparison with V_a, and eqn (2.15) may be written as

$$V_a \approx E_a \qquad (2.24)$$

In the steady state, therefore, the operating conditions of the machine are determined by the fact that the generated EMF (or *back EMF*, as it is usually termed in a motor) must be approximately equal to the armature supply voltage. Since the back EMF E_a is related to the speed and field flux by the equation

$$E_a = K_a \Phi \omega \qquad [2.10]$$

then

$$V_a \approx K_a \Phi \omega$$

DIRECT CURRENT MACHINES

or

$$\omega \approx \frac{V_a}{K_a \Phi} \qquad (2.25)$$

This equation is the basis of motor speed control, for it shows that the speed varies directly with the supply voltage V_a and inversely with the flux Φ. The other quantity of interest is the torque given by

$$T = K_a \Phi I_a \qquad [2.11]$$

If we idealise the machine by putting $\Phi = (K/K_a)I_f$, these equations become

$$\omega \approx \frac{V_a}{KI_f} \qquad (2.26)$$

$$T = KI_f I_a \qquad (2.27)$$

Various methods of connecting the machine impose constraints between V_a, I_f and I_a, and we now explore some of the characteristics that may be obtained in this way.

Separately excited motor

With separate excitation (figure 2.18), I_f and V_a are controlled independently; the speed is given explicitly by eqn (2.26) and the torque by eqn (2.27). A motor is usually designed for a certain nominal speed at nominal values of armature voltage and field current; we examine the effect on the speed of varying the field current I_f and the armature voltage V_a from these nominal values.

Consider first the effect of varying the field current I_f. The normal value of field current will usually take the iron near to magnetic saturation. If the current I_f is increased above its normal value, saturation occurs; there is no longer a linear relationship between Φ and I_f, and eqn (2.26) is not valid. Equation (2.25) shows that the speed varies inversely with the flux Φ, and the flux cannot change appreciably once the iron is saturated. Thus speed control is possible only if the field current is reduced below its normal value − a process known as *field weakening*. This causes the motor speed to rise above its nominal value, as shown by eqn (2.26). The speed range obtainable by field weakening is rather restricted, for I_f cannot be reduced without limit. Equation (2.27) shows that the armature current I_a must increase inversely with I_f to maintain the torque, and commutation difficulties arise when the armature current is large and the field flux small.

Since field control can raise the speed only above the nominal value, armature voltage control must be used for speeds below the nominal value. Equations (2.25) and (2.26) show that the speed is nearly proportional to the armature voltage (for a constant field current), and a very wide speed range may be obtained by varying V_a. This linear relationship between speed and voltage is an important characteristic of the DC machine, which gives it a dominant position in speed control systems (both manual and automatic), with sizes ranging from a few watts to tens of megawatts. Field current and armature voltage control are often combined in applications which require exceptionally large speed ranges.

Permanent-magnet motor

The permanent-magnet motor may be regarded as a separately excited motor with constant field flux. Its speed is nearly proportional to the armature supply voltage, and it has the advantage of higher efficiency because there is no energy loss in a field winding. The gain in efficiency can be significant for small motors. Laithwaite [4] has introduced a 'goodness factor' G for electromagnetic devices which gives a general measure of their performance; for similar devices, a larger value of G implies a higher efficiency. If all the linear dimensions of a device are multiplied by a factor x, then G will be multiplied by x^2. Thus electromagnetic devices get better as they get bigger, and conversely it is difficult to make a small motor with a high efficiency. The losses in the field and armature windings are often comparable in magnitude, so it is advantageous to replace the field winding with a permanent magnet in small motors. In large motors the field-winding loss is only a small fraction of the output power, and other factors such as material cost and manufacturing difficulty preclude the use of permanent magnets.

Shunt motor

In the shunt motor (figure 2.19) the field current I_f is obtained from the armature supply voltage V through a resistance R_f (internal winding resistance plus external variable resistance). Thus $I_f = V/R_f$, and eqns (2.26) and (2.27) become

$$T = \frac{K}{R_f} V I_f \tag{2.28}$$

$$\omega \approx \frac{R_f}{K} \tag{2.29}$$

Thus the speed is independent of V, and varies directly with R_f. Torque and speed are still independent, and the shunt motor is essentially a constant-speed machine. As with the separately excited machine, the range of speed variation obtainable by field weakening (increasing R_f) is limited, and the variable field resistance is normally used only for small variations from the nominal speed. If the effects of armature resistance are included, eqn (2.29) becomes

$$\omega = \frac{R_f}{K} \left\{ 1 - \frac{R_a R_f T}{K V^2} \right\} \tag{2.30}$$

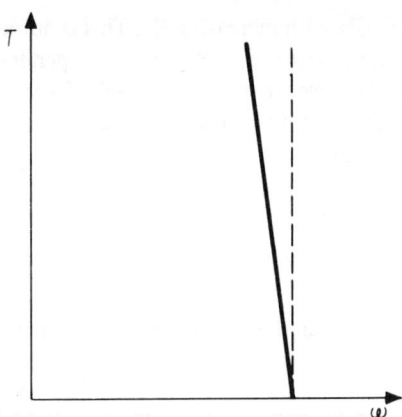

Figure 2.21 Torque/speed characteristic of a shunt motor

and the torque/speed curve is shown in figure 2.21. In practice this form of characteristic is obtained only with small machines which have a relatively large value of R_a. With large machines, the resistive term is small and armature reaction causes the speed to rise with increasing load because the pole flux is reduced.

Some measure of speed control is possible by inserting additional resistance in series with the armature. Variation of this resistance gives a family of torque/speed curves, as shown in figure 2.22. If the torque/speed characteristic of the load is plotted on the same graph, its intersection with the motor torque/speed curve gives the speed at which the motor drives the load. The loaded speed of

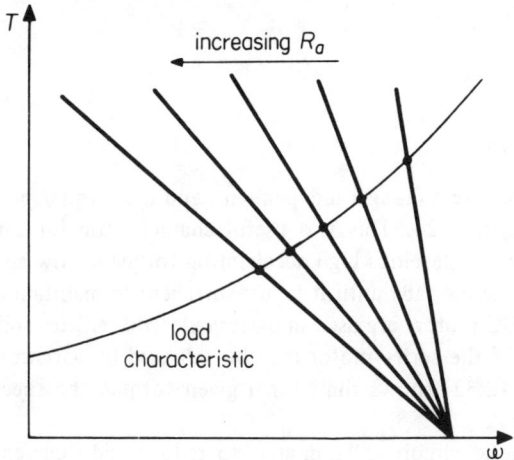

Figure 2.22 Speed control by armature resistance variation

the motor therefore falls with increasing R_a. This is not a good method of speed control, because of the power loss in R_a and the dependence of the speed on the load torque. It does, however, permit the speed of a shunt motor to be reduced below the nominal value; field control, as we have seen, is useful only for raising the speed above this value.

Series motor

If the field winding consists of a few turns capable of carrying the full armature current, it may be connected in series with the armature (figure 2.23). This

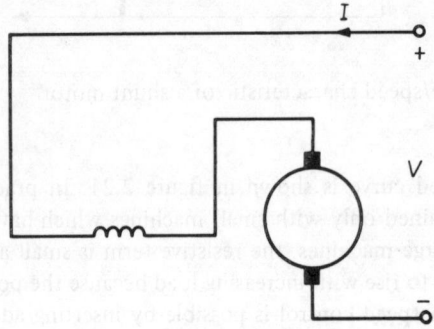

Figure 2.23 Series motor

introduces the constraint $I_f = I_a$ and if resistance is neglected we have, from eqns (2.25) and (2.27)

$$I = \sqrt{(T/K)} \tag{2.31}$$

$$\omega = \frac{V}{\sqrt{(KT)}} \tag{2.32}$$

Torque and speed are no longer independent, and the torque/speed characteristic is shown in figure 2.24. This is a useful characteristic for a vehicle traction motor; the machine develops high accelerating torque at low speeds, and as the speed rises the torque falls until it is just sufficient to maintain the speed of the vehicle. Series DC motors are used in battery electric vehicles and electric trains. Speed control of the series motor may be achieved by variation of the supply voltage V; eqn (2.32) shows that, for a given torque, the speed ω is directly proportional to V.

If the magnetic circuit is laminated to reduce eddy currents, a series DC motor will operate quite well from a single-phase AC supply. Universal motors of this kind are widely used in portable power tools and domestic appliances such

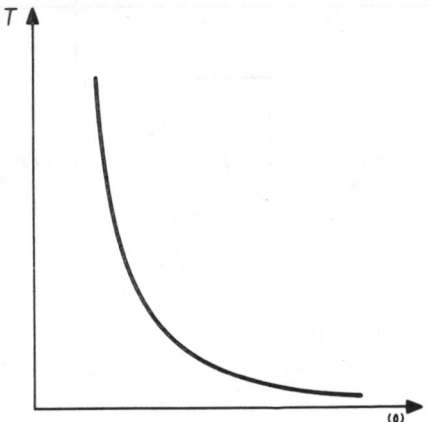

Figure 2.24 Torque/speed characteristic of the series motor

as food mixers. The reason for using the relatively expensive construction of a commutator machine is that much higher speeds are possible than with induction motors, giving a larger power output from a given size. Also, the torque/speed characteristic of the series motor is more suitable than the induction motor characteristic for these applications.

An important characteristic of the series motor is the high speed attained when the machine is lightly loaded; the current in the series field is low, and a high rotational speed is needed to generate the required back EMF in the weak field flux. With a small machine the windage and friction torque is sufficient to limit the no-load speed to a safe value, but a large series motor must never be started without a load or the speed will rise to a very high value and the armature may burst under the rotational stresses. For this reason an auxiliary shunt winding is sometimes added to limit the no-load speed.

Compound motor

A DC motor is sometimes built with both shunt and series field windings connected as shown in figure 2.25. If the series field has many turns, this gives a characteristic in between that of a shunt motor and a series motor, as shown in figure 2.26. Frequently the series field has only a few turns, and its function is to counteract the effects of armature reaction so that the speed does not rise with increasing load.

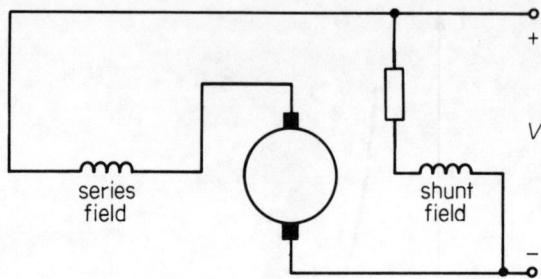

Figure 2.25 Compound motor

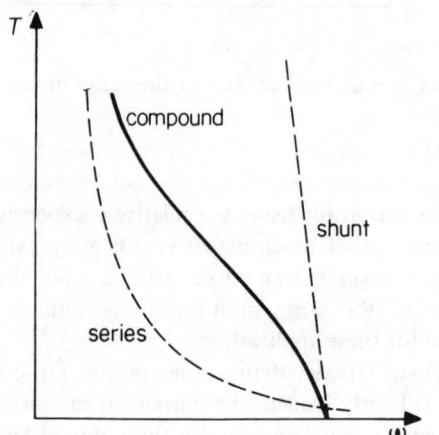

Figure 2.26 Torque/speed characteristic of the compound motor

Electronic control

Control of the speed of a DC motor requires a controllable voltage source; but available power sources – mains or batteries – are essentially fixed voltage. Techniques of power electronics [5] enable a variable voltage to be derived from one of these sources, and the combination of an electronic controller with a DC motor makes a most effective variable-speed drive system. This system has dominated the market for over a decade, but it faces growing competition from the AC variable-frequency system discussed in section 6.4.

Power electronic controllers fall into two main groups [1]: (a) phase control, where a unidirectional voltage is derived from the AC mains by phase-controlled rectification using thyristors; (b) chopper control, where transistors or thyristors switch a DC supply on and off alternately, thereby controlling the mean voltage applied to the motor. Both types of controller allow the mean voltage to be varied over a wide range, with a corresponding range of speed variation. There is, however, a pulsating component of voltage which causes the armature current to fluctuate. With phase control it is possible for the current to be discontinuous; current may cease to flow for a part of each cycle of the AC mains.

There will be additional losses in DC motors operated from power electronic controllers. The field system is usually laminated to reduce eddy-current losses caused by flux pulsations, so the additional loss from this source is small. More serious is the additional I^2R loss in the armature caused by the current fluctuation, particularly if the current is discontinuous. Suppose that the armature current has a mean value I_1 and that the fluctuating component has an RMS value I_2. If the field flux is constant, the mean torque is

$$T = K_a \Phi I_1 \tag{2.33}$$

and the armature I^2R loss is

$$P_a = (I_1^2 + I_2^2) R_a \tag{2.34}$$

Thus, for a given mean torque, the current fluctuation will cause an additional loss equal to $I_2^2 R_a$. In practice the loss may be higher than this because skin effect increases the effective resistance of the armature at the frequencies of the harmonic components of the pulsation. An additional loss of about 10 per cent is common with phase-controlled motors [1].

Starting of DC motors

We have seen that the steady-state operation of DC motors is governed by the condition that the armature back EMF E_a is approximately equal to the supply voltage V_a. This condition does not hold when the motor is started from rest, for E_a is initially zero; if the armature were connected directly to the supply, a large current would flow, limited only by the armature resistance R_a. In fractional-kilowatt motors R_a is usually large enough to limit the starting current to a safe value, and such motors can be started by direct connection to the supply (direct-on-line or DOL starting). But large motors have a relatively small R_a, and a very large current would flow if the armature were connected directly to the full supply voltage. To prevent damage to the motor — particularly the commutator and brushes — the starting current must be limited. Electronic speed controllers normally incorporate a 'soft-start' facility to limit the starting current by control of the armature voltage. Large motors without electronic control are connected to the supply via a starter; this inserts a variable resistor in series with the armature, and the value of the resistor is progressively reduced as the armature runs up to speed.

The field current of a shunt or separately excited motor should always be set to its maximum value for starting, to give maximum starting torque with the available armature current and to give a rapid build-up of the back EMF as the armature accelerates. It is dangerous to attempt to start a motor with the field winding disconnected, because the residual flux may give enough torque to accelerate the armature, which will attain a dangerously high speed in order to generate the required back EMF. For the same reason, the field supply to a motor must never be disconnected while the machine is running; the armature supply must always be switched off first. These precautions do not, of course, apply to series motors, but the danger of starting a series motor without a mechanical load has already been mentioned.

2.6 Dynamic characteristics of DC machines

Dynamic equations

When a DC machine is not operating under steady-state conditions, it is necessary to take account of the inductance of the armature and field circuits and the inertia of the armature. With the circuit of figure 2.27, and the idealised machine equations, we have

$$e_a = Ki_f \omega \tag{2.13}$$

$$T = Ki_f i_a \tag{2.14}$$

$$v_f = R_f I_f + L_f \frac{di_f}{dt} \tag{2.35}$$

$$v_a = e_a + R_a i_a + L_a \frac{di_a}{dt} \tag{2.36}$$

$$T_m = T - J \frac{d\omega}{dt} \tag{2.37}$$

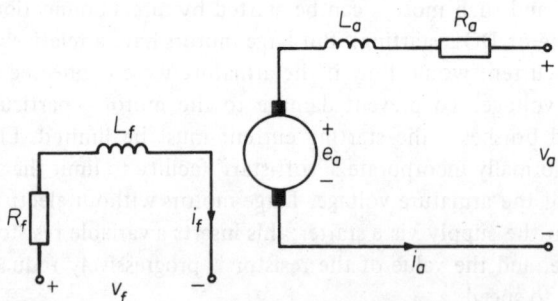

Figure 2.27 DC machine: dynamic circuit model

where T_m is the mechanical output torque and J is the armature moment of inertia.

All quantities may now be time-varying, and the electrical quantities are written with lower-case letters to emphasise this. Note that there is no mutual inductance coupling between the field and armature circuits because the magnetic axes are at right angles; also that the machine is a unilateral device — the field current affects the armature circuit, but the converse is not true. In the mechanical equation of motion, rotational losses have been ignored.

Because of the product terms appearing in eqns (2.13) and (2.14), specific problems often yield non-linear differential equations which are not soluble analytically; numerical methods must then be used. In many applications, however, one of the quantities in the product is a constant, and the solution of the problem is quite straightforward. Two simple examples will illustrate the method.

Transient performance of DC generators

Consider a DC generator driven at a constant speed ω, with the armature on open circuit. Suppose that a steady voltage V_f is suddenly applied at time $t = 0$ to the field winding; the field current is given by eqn (2.35), and the solution is

$$i_f = \frac{V_f}{R_f} (1 - e^{-t/\tau_f}) \tag{2.38}$$

where $\tau_f = L_f/R_f$ is the field time constant. The generated EMF, given by eqn (2.14), is then

$$e_a = K\omega i_f = \frac{K\omega V_f}{R_f} (1 - e^{-t/\tau_f}) \tag{2.39}$$

After a sufficient length of time this will attain the steady value

$$E_a = \frac{K\omega V_f}{R_f} \tag{2.40}$$

Suppose now that a steady voltage V_a is suddenly applied to the armature terminals; the armature current is given by eqn (2.36), with the solution

$$i_a = \frac{V_a - E_a}{R_a} (1 - e^{-t/\tau_a}) \tag{2.41}$$

where $\tau_a = L_a/R_a$ is the armature time constant. Note that there is no reflected voltage in the field circuit due to current in the armature circuit.

Starting of DC motors

As a second example of DC machine dynamics, consider the starting process of a shunt or separately excited motor. We assume a constant field current I_f, and no mechanical load so that $T_m = 0$. At time $t = 0$, a constant supply voltage V_a is suddenly connected to the armature. If the total armature circuit resistance is R_a, and the inductance may be neglected, then

$$V_a = e_a + R_a i_a$$
$$= K\omega I_f + R_a i_a \qquad (2.42)$$

Also

$$J \frac{d\omega}{dt} = T = K I_f i_a \qquad (2.43)$$

so that

$$V_a = K I_f \omega + \frac{R_a J}{K I_f} \frac{d\omega}{dt} \qquad (2.44)$$

The solution to this equation is

$$\omega = \Omega(1 - e^{-t/\tau_{em}}) \qquad (2.45)$$

where $\Omega = V/KI_f$ is the final steady-state speed and the time constant τ_{em} is given by

$$\tau_{em} = \frac{R_a J}{(KI_f)^2} \qquad (2.46)$$

This quantity is termed the *electromechanical time constant* and it is generally much longer than the armature time constant τ_a. If the armature inductance L_a is not neglected, the starting process is described by the second-order differential equation

$$V_a = \frac{L_a i_a}{K I_f} \frac{d^2\omega}{dt^2} + \frac{R_a J}{K I_f} \frac{d\omega}{dt} + K I_f \omega \qquad (2.47)$$

Provided that $\tau_{em} \gg \tau_a$, the solution to this equation may be divided into two parts [6]

$$t \ll \tau_{em}; \quad \omega \approx 0, \, i_a = \frac{V_a}{R_a}(1 - e^{-t/\tau_a}) \qquad (2.48)$$

$$t \gg \tau_a; \quad \omega \approx \Omega(1 - e^{-t/\tau_{em}}), \, i_a = \frac{V_a - K I_f \omega}{R} \qquad (2.49)$$

Equation (2.48) describes an electrical transient, where the armature current rises exponentially towards a limiting value V_a/R_a before the armature has time to accelerate. Equation (2.49) describes a mechanical transient, where the armature speed rises exponentially towards its final limiting value Ω, and the armature current is limited only by the resistance R_a; the current is changing too slowly for the inductance L_a to have any significant effect. Thus the electrical and mechanical transients can be treated independently, provided that the two time constants are widely different. This is an instance of an important separation principle, which applies to many transient problems in electrical machines.

2.7 Special machines

The majority of DC machines have the same basic form, regardless of size: the armature conductors lie in slots in a cylindrical iron rotor, and the field system — whether permanent magnet or otherwise — is stationary. DC machines which do not conform to this pattern have been developed for special purposes. Only the more common forms of special machine are described in this section; a comprehensive account will be found in Say and Taylor [1].

Moving-coil machines

Motors and tachogenerators used in control systems are often required to have a very low rotor inertia so that the system can respond rapidly. With small machines this can be achieved by detaching the armature conductors from the iron core; the conductors are bonded in resin to form a 'basket' which is free to rotate in the airgap between the field poles and a fixed iron cylinder. Moving-coil motors of this kind generally develop a very smooth torque, since there are no rotor slots to cause 'cogging'. By the same token, moving-coil tachogenerators develop a smooth EMF. Low inertia is achieved at the expense of a large airgap, with a correspondingly large MMF to be supplied by the field windings or permanent magnets.

Disc machines

In a conventional DC machine, the magnetic field is radial and the current flow is axial; the two interact to give a circumferential force which exerts a torque on the cylindrical armature. A torque will also be produced if the directions of field and current are interchanged; this is the principle of the disc machine. An axial field is set up by permanent magnets, usually in opposing pairs. A disc-shaped armature, with conductors arranged radially like spokes, rotates in the axial magnetic field. The disc machine is thus a variant of the moving-coil machine, and it has similar properties. In addition it is very compact, and applications include drives for computer tape decks and engine cooling fans.

The armature of a disc machine can be made from resin-bonded wire, as in the moving-coil machines. In small sizes another technique is widely used: conductors are formed on the surface of an insulated disc, either from punched metal sheet or by the methods used to make printed-circuit boards. For this reason disc motors are often called *printed-armature motors*. The construction of a typical printed-armature motor is shown in figure 2.28.

Figure 2.28 Printed-armature DC motor (Printed Motors Ltd)

Linear DC motors

The disc machine is derived from the ordinary cylindrical machine by interchanging the directions of field and current; the force is still circumferential, giving rotary motion. Suppose that the radial field is retained, but the current direction is made circumferential; the force will then be axial, giving linear motion. This is the principle of the moving-coil loudspeaker mentioned in section 1.2, which may be regarded as a homopolar linear motor with very limited movement. A simple modification will allow much greater movement: the coil is wound on a long iron bar; current is supplied to a small section of this coil through brushes, and a radial magnetic field is applied to the energised section of the coil between the brushes. As the bar moves, the brushes continually transfer the current to a new active section of the coil. Although the principle is simple, practical difficulties limit the usefulness of this type of machine. AC linear motors are more widely used than DC, and these are described in sections 5.4 and 6.7.

Brushless DC motors

The commutator is merely a set of mechanical switches, so it is possible to replace it with semiconductor switches. This is the basis of the brushless DC motor. Since it is simpler to have the switches stationary, the functions of stator and rotor are interchanged; the rotor carries a permanent-magnet field system, and the armature coils are on the stator. Semiconductor switches — usually transistors — must reverse the currents in the armature coils at the correct instants as the field poles rotate; it is thus necessary to sense the rotor position and use this to control the switches. We now have a machine in which the stator coils carry alternating currents, and these currents produce a magnetic field which rotates in synchronism with the rotor. This is nothing other than a synchronous machine, which differs from the conventional synchronous machine of chapter 5 in two respects: the stator currents are not sinusoidal, and the frequency is not fixed but is determined by the rotor speed. The term 'brushless DC motor' is therefore a misnomer, but it has become established. These machines have the useful characteristics of DC motors with the added advantages of high reliability and the absence of sparking. They are increasingly used in applications such as computer disc drives where high reliability and consistent performance are important.

Problems

2.1. If the rotational losses of a DC shunt motor are constant, prove that the efficiency of the motor will be a maximum when the armature current is such that the armature resistance loss $I_a^2 R_a$ is equal to the sum of the rotational loss and the field resistance loss.

A 500 V DC shunt motor has a field winding resistance of 1000 Ω and a rated full-load output power of 10 kW. If the rotational losses amount to 250 W and the efficiency is a maximum at full load, calculate this efficiency and the value of the armature resistance.

2.2. If the field and armature windings of a DC machine carry alternating currents with RMS values I_f and I_a respectively, show that the average torque developed by the machine is given by

$T = K I_f I_a \cos \phi$

where ϕ is the phase angle between the currents and K is the machine constant.

A series motor is connected to an alternating voltage supply of RMS value V. Show that the average torque is given by

$$T = \frac{KV^2}{(R + K\omega_r)^2 + X^2}$$

where K is the machine constant, ω_r is the armature angular velocity, R is the total series resistance and X is the total series reactance of the windings at the frequency of the supply. Obtain the corresponding torque expression for the same motor operating from a DC supply and discuss the difference between AC and DC operation.

2.3. The hoisting cable of a crane is wound on to a drum, and a DC shunt motor drives the drum through a reduction gearbox. The crane is used to lift a load at a steady speed, and the speed is controlled by varying a resistance R in series with the armature. The internal armature resistance may be neglected.

Show that the hoisting speed u is given by the expression

$$u = u_0 - AWR$$

where u_0 is the no-load speed, W is the weight of the load and A is a constant. Also show that the efficiency of the system is u/u_0. Rotational losses in the motor and the gearbox may be neglected, and the cable winds on to the drum at a constant radius.

2.4. A 200 V DC shunt motor has a no-load speed of 100 rad/s and its armature resistance is 1 Ω. Rotational losses are negligible. The motor drives a water pump, and the normal torque load on the motor is 40 N m. A fault in the water system causes the torque load on the motor to fall suddenly to 20 N m. Calculate

(a) the normal armature current
(b) the normal speed at which the motor drives the pump
(c) the motor armature current just after the fault has occurred
(d) the final armature current
(e) the final motor speed.

If the rotating parts have a moment of inertia of 0.1 kg m^2, and the inductance of the armature may be neglected, obtain the differential equation which governs the speed of the motor after the fault.

2.5. A separately excited DC motor has a constant field current I_f. When a voltage v is applied to the armature a current i will flow; if the motor is unloaded, the resulting torque will accelerate the armature. Neglecting rotational losses, show that

$$v = \frac{(KI_f)^2}{J} \int i \, dt + R_a i + L_a \frac{di}{dt}$$

where J is the moment of inertia and K is the machine constant. Hence show that the machine is equivalent to a series combination of resistance R_a, inductance L_a and capacitance $C_m = J/(KI_f)^2$. If a constant voltage V is applied for a sufficiently long time, the current will be zero and the energy stored in the capacitance will be $\tfrac{1}{2}C_m V^2$. Prove that this is equal to the rotational energy of the armature.

References

1 M. G. Say and E. O. Taylor, *Direct Current Machines* (London: Pitman, 1980).
2 G. W. Carter, *The Electromagnetic Field in its Engineering Aspects*, 2nd ed. (London: Longman, 1967).
3 B. Hague, *The Principles of Electromagnetism Applied to Electrical Machines* (New York: Dover, 1962).
4 E. R. Laithwaite, 'The goodness of a machine', *Proc. IEE*, **112** (1965), pp. 538-41.
5 S. B. Dewan and A. Straughen, *Power Semiconductor Circuits* (New York: Wiley, 1975).
6 H. Majmudar, *Introduction to Electrical Machines* (Boston: Allyn and Bacon, 1969).

3 Alternating Current Systems

3.1 Introduction

With DC machines, the voltage generated by an individual armature coil is an alternating quantity which is rectified mechanically by the commutator. If we dispense with the commutator and revert to the slipring model of figure 2.1, we have a rudimentary AC generator. Practical AC generators are essentially simpler than their DC counterparts and are more easily designed in very large sizes; but a more important reason for their adoption is the possibility of using transformers to raise the voltage level for power transmission over long distances, and then to reduce it again for domestic or industrial consumption.

In nearly all applications, alternating current has advantages over direct current. The usefulness of the transformer is one reason, but it is the induction motor more than any other device which has vindicated the alternating current system. Induction motors are cheap, efficient and robust; they supply most of the motive power for industry, and they are used in many domestic appliances. The other main type of AC machine is the synchronous machine; most AC generators are of this kind, and in large sizes the synchronous motor is a strong rival to the induction motor. These machines are treated in chapters 4, 5 and 6, but it is first necessary to consider some of the general properties of AC systems; this is done in sections 3.2 and 3.3.

Transformers are considered in section 3.4. Although they have no moving parts they are traditionally included with electrical machines, for the theory of the transformer makes a useful introduction to the theory of AC rotating machines. The idea of an equivalent circuit, which arises naturally in the study of the transformer, proves to be a useful concept in rotating machines. Imperfections such as magnetic leakage and core loss are present in AC machines as well as transformers, and they can be represented by similar elements in the equivalent circuits.

3.2 Generation of sinusoidal alternating voltages

The EMF generated in a coil depends only on the rate of change of the flux linking the coil; it is of no consequence whether the coil moves in the field of fixed magnetic poles, or the poles move and the coil remains stationary. Usually, it is more convenient to have the active coils stationary and field poles rotating, as shown in the simple model of figure 3.1. This is always done with large machines, to avoid the transfer of large amounts of power through sliprings. In the simple model, the rotor is magnetised by a field coil (or 'excitation winding'), with the field current supplied via brushes and sliprings (not shown in figure 3.1). The stationary part of the machine (the 'stator') carries a single turn armature coil, and an EMF will be induced in this coil when the rotor moves. If the rotor angular velocity is ω, the induced EMF, which has already been calculated for the DC machine, is

$$e = 2Blr\omega \tag{2.1}$$

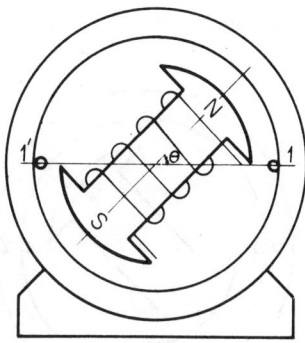

Figure 3.1 Simple single-phase AC generator

where l is the length of a coil side and r is the radius. The flux density B is the value at the right-hand coil side, which is displaced by an angle θ from the magnetic axis of the rotor. During each revolution of the rotor B will vary with θ, and by suitably shaping the poles it can be arranged that

$$B = B_m \cos \theta \tag{3.1}$$

If θ has some value θ_0 at time $t = 0$, and the angular velocity ω is constant, then $\theta = \omega t + \theta_0$ and eqn (3.1) becomes

$$B = B_m \cos(\omega t + \theta_0) \tag{3.2}$$

The induced EMF is now a sinusoidal alternating quantity, given by

$$\begin{aligned} e &= 2lr\omega B_m \cos(\omega t + \theta_0) \\ &= E_m \cos(\omega t + \theta_0) \end{aligned} \tag{3.3}$$

where $E_m = 2lr\omega B_m$. Strictly speaking this is a cosinusoidal quantity; but the term 'sinusoidal' will be used to describe a sine or a cosine function. Sine waves are distinguished from all other periodic functions by the fact that the steady-state response of any linear electric circuit to a sine wave of voltage is also sinusoidal. It is therefore advantageous to generate alternating voltages in the form of sine waves for general transmission and distribution, and generators are normally designed to do this. As with DC machines, practical AC generators differ from the simple model in having many armature coils; these machines are considered in chapter 4.

3.3 Polyphase systems

Suppose that a second armature coil is added to the simple generator, at right angles to the first, as shown in figure 3.2. If the flux density at coil side α is

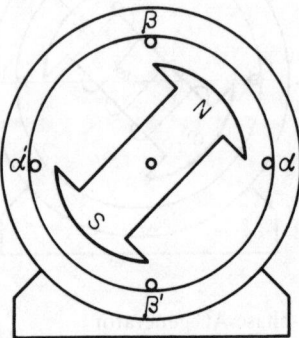

Figure 3.2 Simple two-phase AC generator

$B_m \cos \theta$, the corresponding value at coil side β is $B_m \cos (\theta - \pi/2)$. The generated voltages are then given by

$$e_\alpha = E_m \cos (\omega t + \theta_0)$$
$$e_\beta = E_m \cos (\omega t + \theta_0 - \pi/2)$$
(3.4)

These voltages have the same frequency but different phase angles; the two armature coils are known as *phases*, and the voltages are the *phase voltages*. This two-phase system is a rather special case, and the general symmetrical m-phase (or polyphase) system is obtained from m coils arranged symmetrically round the armature. Thus for a three-phase generator (figure 3.3) we have

$$e_a = E_m \cos (\omega t + \theta_0)$$
$$e_b = E_m \cos (\omega t + \theta_0 - 2\pi/3)$$
$$e_c = E_m \cos (\omega t + \theta_0 - 4\pi/3)$$
(3.5)

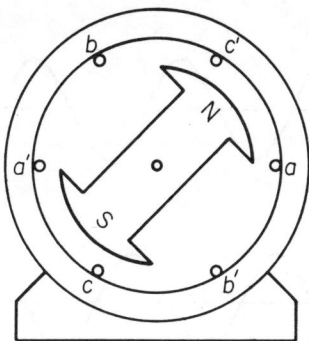

Figure 3.3 Simple three-phase AC generator

These voltages may be represented by the phasor diagram of figure 3.4, and the corresponding waveforms are shown in figure 3.5. The generator is said to be *balanced* when all the phase voltages have the same amplitude, as they do here.

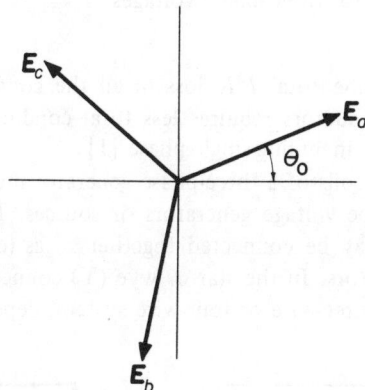

Figure 3.4 Phasor diagram for three-phase voltages

Three-phase systems

Most AC machines require balanced polyphase currents for satisfactory operation. (A notable exception is the small single-phase induction motor, discussed in section 6.6.) In principle any number of phases upwards of two could be used, with machines of appropriate design; in practice three phases are almost universally used for economic reasons. The economics of AC power transmission systems may be compared on the basis of the same maximum voltage between

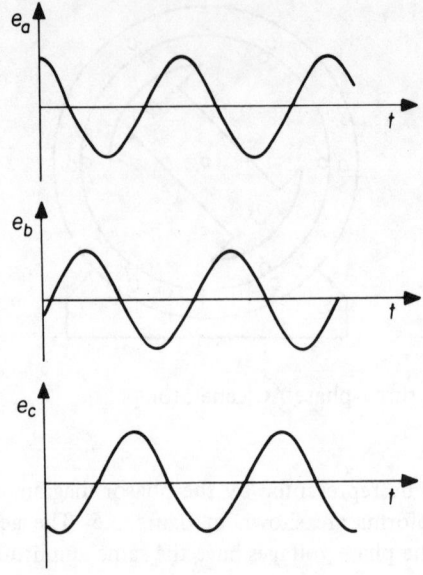

Figure 3.5 Waveforms for three-phase voltages

conductors and the same total I^2R loss in all the conductors; a three-phase system using three conductors requires less total conductor material than any other number of phases, including single phase [1].

The three armature coils of a three-phase generator may be represented in a circuit diagram by three voltage generators or sources. There are two ways in which these sources may be connected together so as to transmit power with fewer than six conductors. In the star or wye (Y) connection of figure 3.6(a), there is a choice of a three-wire or four-wire system, depending on whether the

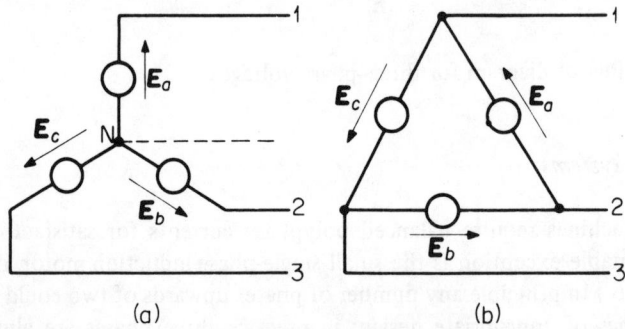

Figure 3.6 Connection of a three-phase source: (a) star, (b) delta

star (or neutral) point N is connected to a fourth line. The mesh or delta (Δ) connection of figure 3.6(b) is possible in a balanced system because $e_a + e_b + e_c = 0$ at all instants of time (this may be seen from the phasor diagram; the sum of the three phasors is zero). Only a three-wire system is possible with a delta-connected source. In each case the individual sources are known as *phases*. If we consider only three-wire systems, a three-phase load may be connected to the lines in two ways (figure 3.7). The impedances Z_a, Z_b and Z_c form the phases of the load; if they are equal the load is balanced.

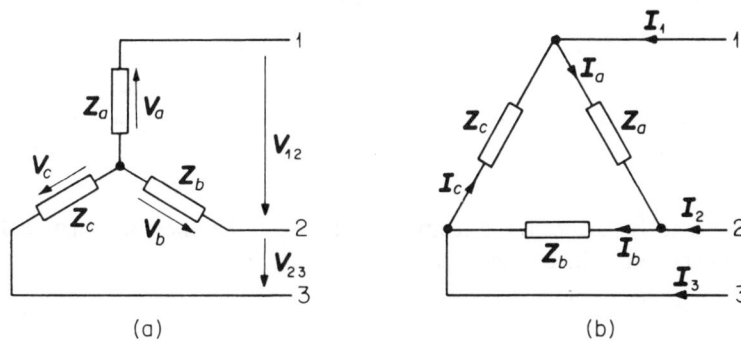

Figure 3.7 Connection of a three-phase load: (a) star, (b) delta

When the load is star connected (figure 3.7(a)), the current in each phase of the load is equal to the current in the corresponding line. The voltage across a phase, however, is not equal to the voltage between a pair of lines. When the source and load are balanced, the voltage phasor diagram takes the symmetrical form shown in figure 3.8(a), from which we obtain the relationship

$$V_{\text{line}} = \sqrt{3}\, V_{\text{phase}} \tag{3.6}$$

With a delta-connected load (figure 3.7(b)), the voltage across each phase is equal to the voltage between the corresponding pair of lines. It is now the phase and line currents which are unequal, and the current phasor diagram for a balanced system takes the form of figure 3.8(b); this gives the relationship

$$I_{\text{line}} = \sqrt{3}\, I_{\text{phase}} \tag{3.7}$$

It will be seen that star and delta connection are duals, with voltage in one analogous to current in the other. Unbalanced systems may be handled by the usual techniques of circuit analysis, with voltage or current sources representing the phases of the generator. The star–delta (T-π) transformation of circuit theory is often useful for converting star loads to equivalent delta form, and vice versa.

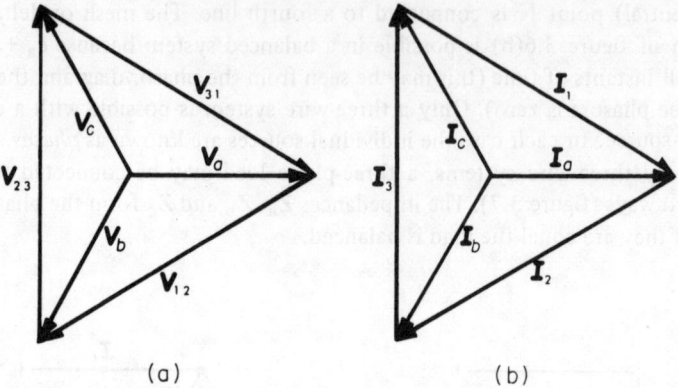

Figure 3.8 Phasor diagrams for a balanced load: (a) star, (b) delta

When the load is balanced, the relationship between the phase impedances for equivalent star and delta configurations is

$$Z_{\text{delta}} = 3 Z_{\text{star}} \tag{3.8}$$

Phase sequence

In the three-phase system so far considered, the voltages pass through their maximum positive values in the sequence a → b → c. This is the normal arrangement, and it is termed the *positive phase sequence*. If any pair of phases is interchanged the sequence will be reversed; thus interchanging e_b and e_c will give the system

$$e_{a'} = e_a = E_m \cos(\omega t + \theta_0)$$

$$e_{b'} = e_c = E_m \cos(\omega t + \theta_0 - 4\pi/3) = E_m \cos(\omega t + \theta_0 + 2\pi/3)$$

$$e_{c'} = e_b = E_m \cos(\omega t + \theta_0 - 2\pi/3) = E_m \cos(\omega t + \theta_0 + 4\pi/3)$$

The phase sequence for this system is c' → b' → a', which is termed the *negative sequence*. It will be shown in chapter 4 that the direction of rotation of an AC motor depends on the phase sequence of the supply, and a positive sequence is normally assumed.

3.4 Transformers

A transformer is a pair of coils coupled magnetically (figure 3.9), so that some of the magnetic flux produced by the current in the first coil links the turns of the

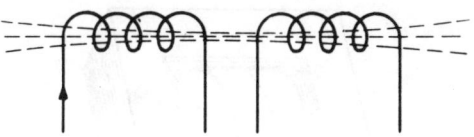

Figure 3.9 Coupled coils

second, and vice versa. The coupling can be improved by winding the coils on a common magnetic core (figure 3.10), and the coils are then known as the 'windings' of the transformer. Practical transformers are not usually made with the windings widely separated as shown in figure 3.10, because the coupling is not very good. Exceptionally, some small power transformers, such as domestic bell transformers, are sometimes made this way; the physical separation allows the coils to be well insulated for safety reasons. Figure 3.11 shows the shell type of construction which is widely used for single-phase transformers. The windings are placed on the centre limb either side-by-side or one over the other, and the magnetic circuit is completed by the two outer limbs. Say [2] gives details of different types of transformer construction and of three-phase transformers, which are beyond the scope of this book.

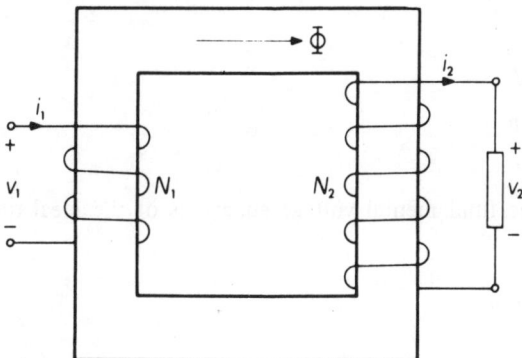

Figure 3.10 Elementary transformer

The ideal transformer

In an ideal transformer, the same core flux Φ links each turn of each winding. Suppose that one winding (known as the *primary*) is connected to a voltage source v_1; it will draw a current i_1, and the voltage equation is

$$v_1 = R_1 i_1 + N_1 \frac{d\Phi}{dt} \qquad (3.9)$$

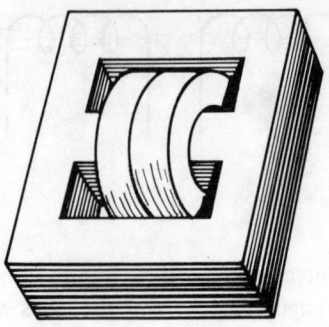

Figure 3.11 Shell-type transformer construction

If the other winding (known as the *secondary*) is connected to a load, a current i_2 will flow, and the terminal voltage v_2 is given by

$$v_2 = N_2 \frac{d\Phi}{dt} - R_2 i_2 \qquad (3.10)$$

the negative sign arising from the different direction of current flow. If the windings have negligible resistance, then eqns (3.9) and (3.10) reduce to

$$v_1 = N_1 \frac{d\Phi}{dt} \qquad (3.11)$$

$$v_2 = N_2 \frac{d\Phi}{dt} \qquad (3.12)$$

and these are the fundamental voltage equations of the ideal transformer. By division

$$\frac{v_2}{v_1} = \frac{N_2}{N_1} = n \qquad (3.13)$$

where n is the turns ratio.

To find a relationship between i_1 and i_2, consider the magnetic circuit of the core. If S is the reluctance, then

$$S\Phi = F = N_1 i_1 - N_2 i_2 \qquad (3.14)$$

The reluctance of the core is given by

$$S = \frac{l}{\mu_0 \mu_r A} \qquad [1.75]$$

and if the relative permeability μ_r is high, S will be small. In an ideal transformer we postulate infinite permeability and therefore zero reluctance; eqn (3.14) becomes

$$N_1 i_1 = N_2 i_2 \qquad (3.15)$$

that is, the primary MMF must balance the secondary MMF. The required current relationship is therefore

$$\frac{i_2}{i_1} = \frac{N_1}{N_2} = \frac{1}{n} \qquad (3.16)$$

showing that the current transformation is the inverse of the voltage transformation. Equations (3.13) and (3.16) may be rewritten in the form

$$v_2 = n v_1 \qquad (3.17)$$
$$i_2 = \frac{1}{n} i_1 \qquad (3.18)$$

Multiplication of eqn (3.17) by (3.18) gives

$$v_2 i_2 = v_1 i_1 \qquad (3.19)$$

showing that the instantaneous power output is equal to the instantaneous power input.

With steady AC conditions we have the phasor equations

$$V_2 = n V_1 \qquad (3.20)$$
$$I_2 = \frac{1}{n} I_1 \qquad (3.21)$$

If an impedance Z_2 is connected to the secondary (figure 3.12), then $V_2/V_1 = Z_2$. Division of eqn (3.20) by (3.21) gives

$$Z_2 = \frac{V_2}{I_2} = n^2 \frac{V_1}{I_1}$$

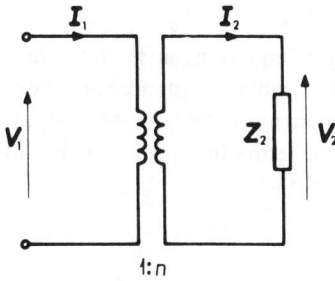

Figure 3.12 Ideal transformer with a load

Now V_1/I_1 is the impedance Z_1 presented by the primary terminals; hence

$$Z_1 = \frac{1}{n^2} Z_2 \tag{3.22}$$

so the transformer has the property of changing impedances. It is this property which makes the transformer so useful in electronic and communication circuits; in power circuits it is the voltage or current transforming property which is of interest.

The real transformer

In a real (as opposed to an ideal) transformer, the winding resistances are not zero; the magnetic coupling between the coils is not perfect; and the reluctance of the core is not zero. We now show that it is possible to represent the real transformer by an equivalent circuit consisting of an ideal transformer together with other elements which represent the imperfections. Since the transformer is a pair of coupled coils, we may use the coupled circuit equations derived in section 1.3, with the notation of figure 3.13

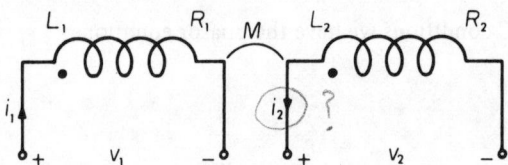

Figure 3.13 Coupled coils: notation

$$\left. \begin{array}{l} v_1 = R_1 i_1 + L_1 \dfrac{di_1}{dt} - M \dfrac{di_2}{dt} \\[6pt] v_2 = -R_2 i_2 - L_2 \dfrac{di_2}{dt} + M \dfrac{di_1}{dt} \end{array} \right\} \tag{3.23}$$

The negative signs in these equations arise from the reversed direction of the secondary current i_2. The winding resistances occur explicitly in these equations but it is not obvious how the magnetic imperfections are included. It is necessary to rearrange the inductive terms in eqns (3.23) by introducing the concept of leakage inductance.

Leakage inductance

The reluctance of the magnetic circuit is finite, and the magnetic circuit law $NI = S\Phi$ implies that a current is required to set up the working flux in the transformer core. Thus if a current i_1 flows in the primary winding, with no current in the secondary, the flux linkage with the primary winding is given by

$$\psi_1 = L_1 i_1 \tag{3.24}$$

and the average flux per turn is

$$\Phi_1 = \frac{L_1 i_1}{N_1} \tag{3.25}$$

Some of the flux produced by the primary current will also link the secondary winding; the flux linkage is given by

$$\psi_{21} = M i_1 \tag{3.26}$$

and the average flux per turn for the secondary is

$$\Phi_{21} = \frac{M i_1}{N_2} \tag{3.27}$$

If the coupling were perfect the same flux would link each turn of each winding. On account of magnetic leakage (figure 3.14) the flux Φ_{21} linking the turns of

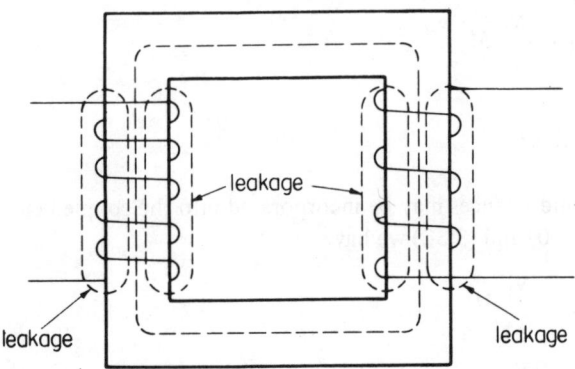

Figure 3.14 Leakage flux

the secondary will be less than the flux Φ_1 linking the primary, and we may define the leakage flux Φ_{l_1} as the difference between these quantities

$$\Phi_{l_1} = \Phi_1 - \Phi_{21} \tag{3.28}$$

Since Φ_1 and Φ_{21} are both proportional to the current i_1 we may put

$$\Phi_{l_1} = \frac{l_1 i_1}{N_1} \tag{3.29}$$

The quantity l_1 is termed the *leakage inductance* of the primary winding, and its physical interpretation is as follows: the flux produced by the current i_1 flowing in the inductance l_1 represents that portion of the primary flux which fails to link with the secondary.

The leakage inductance l_1 may be expressed in terms of L_1 and M, for we have

$$\Phi_{l_1} = \Phi_1 - \Phi_{21}$$

$$= \frac{L_1 i_1}{N_1} - \frac{M i_1}{N_2} \tag{3.28}$$

and eqn (3.29) gives the result

$$l_1 = \frac{N_1 \Phi_{l_1}}{i_1} = L_1 - \frac{N_1}{N_2} M \tag{3.30}$$

Similarly there is a secondary leakage flux given by

$$\Phi_{l_2} = \Phi_2 - \Phi_{12} = \frac{L_2 i_2}{N_2} - \frac{M i_2}{N_1} \tag{3.31}$$

and a secondary leakage inductance

$$l_2 = L_2 - \frac{N_2}{N_1} M \tag{3.32}$$

Equivalent circuit

The leakage inductances may be incorporated into the coupled circuit equations; from eqns (3.30) and (3.32) we have

$$\begin{aligned} L_1 &= l_1 + \frac{N_1}{N_2} M \\ L_2 &= l_2 + \frac{N_2}{N_1} M \end{aligned} \tag{3.33}$$

and eqns (3.23) become

$$\left. \begin{aligned} v_1 &= R_1 i_1 + l_1 \frac{di_1}{dt} + \frac{1}{n} M \frac{d}{dt}(i_1 - n i_2) \\ v_2 &= -R_2 i_2 - l_2 \frac{di_2}{dt} + M \frac{d}{dt}(i_1 - n i_2) \end{aligned} \right\} \tag{3.34}$$

where $n = N_2/N_1$ as before. Equations (3.34) are the equations of the circuit shown in figure 3.15, which is the required equivalent circuit. This is a time-

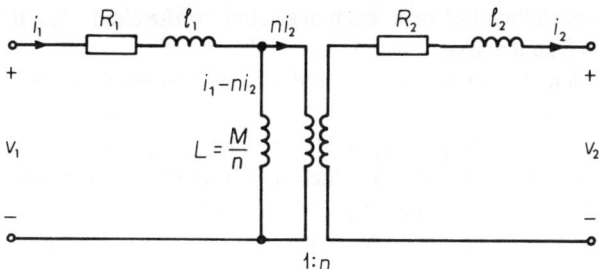

Figure 3.15 Time-domain equivalent circuit of the transformer

domain circuit, valid for arbitrarily time-varying voltages and currents, subject only to the assumptions made in deriving the equations. These are

(a) capacitance effects, between turns and between the two windings, are negligible
(b) magnetic non-linearity is ignored
(c) iron losses are ignored.

We now consider steady-state sinusoidal operation and the representation of iron losses.

AC equivalent circuit of the transformer

For sinusoidal AC operation we may draw an AC (frequency-domain) circuit corresponding to figure 3.15, with inductances replaced by reactances and instantaneous quantities replaced by phasor (complex) quantities. The AC circuit is shown in figure 3.16, in which an additional resistance R_c has been connected in parallel with the reactance jX_m. This resistance represents the iron loss in the core, and the reason for placing it in parallel with jX_m will be explained

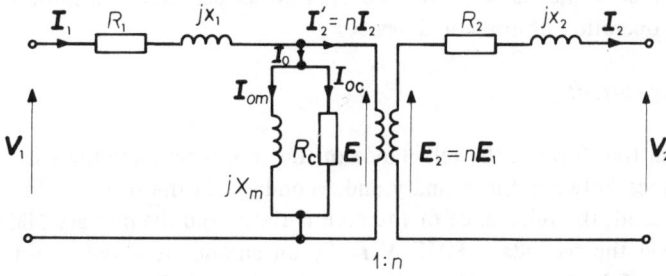

Figure 3.16 Frequency-domain equivalent circuit of the transformer

presently. Capacitance has again been neglected, so the circuit is valid only up to the low audio frequencies.

The equivalent circuit consists of an ideal transformer of ratio $1 : n = N_1 : N_2$, together with elements which represent the imperfections of the real transformer. The voltage E_1 across the primary of the ideal transformer represents the voltage induced in the primary winding by the mutal flux Φ. This is the portion of the core flux which links both primary and secondary coils; the leakage flux associated with one winding does not link the other, and it is represented in the circuit by the leakage reactances jx_1 and jx_2. Similarly, the voltage E_2 across the secondary of the ideal transformer represents the voltage induced in the secondary winding by the mutual flux Φ.

Voltage equation

From eqn (3.11) for the ideal transformer, we have

$$E_1 = j\omega N_1 \Phi \tag{3.35}$$

showing that the induced voltage is in quadrature with the mutual flux. If Φ_{max} is the peak value of the core flux, then the RMS magnitude of the primary voltage is

$$E_1 = \frac{\omega}{\sqrt{2}} N_1 \Phi_{max}$$

$$= 4.44 f N_1 \Phi_{max} \tag{3.36}$$

where f is the frequency of the supply. Equation (3.36) is a basic design equation for a transformer or an iron-cored inductor. The peak core flux is given by

$$\Phi_{max} = A B_{max} \tag{3.37}$$

where A is the cross-sectional area of the core and B_{max} is the peak flux density. The maximum value of B_{max} is set by saturation of the magnetic material, and this fixes the maximum value of Φ_{max} for a given device. Equation (3.36) shows that E_1 must vary in proportion to f if the magnitude of the flux is to remain constant; this is the basis of the 'constant volts per hertz' rule for variable-frequency operation of iron-cored devices.

Magnetising current

In the ideal transformer, the reluctance of the core is zero and there is an exact MMF balance between the primary and secondary. In the real transformer, on the other hand, the reluctance of the core is finite, and the primary MMF $N_1 I_1$ must exceed the secondary MMF $N_2 I_2$ by an amount required to set up the mutual flux Φ in the core. When the secondary current I_2 is zero, the primary current has a finite value I_0, known as the *no-load current*. This current has a component I_{oc}, in phase with E, to supply the eddy-current and hysteresis losses in the core; and a component I_{om}, known as the *magnetising current*, to set up the core flux. The magnetising current is, of course, the current flowing in the inductive reactance of the primary winding. From figure 3.16 and eqn (3.35) we have

$$I_{om} = \frac{E_1}{jX_m} = \frac{\omega N_1}{X_m}\Phi \qquad (3.38)$$

showing that the magnetising current I_{om} is in phase with the mutual flux Φ.

Core loss

It only remains to justify the use of the resistance R_c to represent losses in the iron core. From eqns (3.36) and (3.37) we have

$$E_1 \propto fB_{max}$$

so the core loss is represented by

$$P_c = E_1^2/R_c \propto f^2 B_{max}^2$$

But from eqns (1.64) and (1.65) we also have

$$P_e \propto f^2 B_{max}^2$$

$$P_h \propto fB_{max}^{1.7}$$

where P_e is the eddy-current loss and P_h is the hysteresis loss. It follows that eddy-current loss can be represented by a constant resistance, but that hysteresis loss must be represented by a resistance which varies with f and B_{max}. Thus the value of R_c is independent of the frequency and voltage only if there is no hysteresis loss; since this is not the case, the correct value of R_c must be chosen for the particular operating conditions. With power transformers the frequency is fixed and E_1 does not vary by more than a few per cent when the primary supply voltage V_1 is held constant, so the variation in the value of R_c may be ignored.

Phasor diagram

When the secondary current I_2 is zero (the secondary is on open circuit) the primary current I_1 is just equal to I_0 — the total current flowing in the shunt elements R_c and jX_m. When a load is connected to the secondary, a current I_2 flows, and the primary current is increased by an amount $I_2' = nI_2$. This is termed the *load component* of the primary current, or the secondary current *referred* to the primary, and the total primary current is given by

$$I_1 = I_0 + nI_2 \qquad (3.39)$$

From the equivalent circuit, we also have

$$V_2 = E_2 - I_2(R_2 + jx_2)$$
$$= nV_1 - nI_1(R_1 + jx_1) - I_2(R_2 + jx_2) \tag{3.40}$$

These and other relationships are conveniently illustrated by the phasor diagram of figure 3.17. Equations (3.39) and (3.30) reduce to the ideal transformer equations when both the no-load current I_0 is zero and the leakage impedances

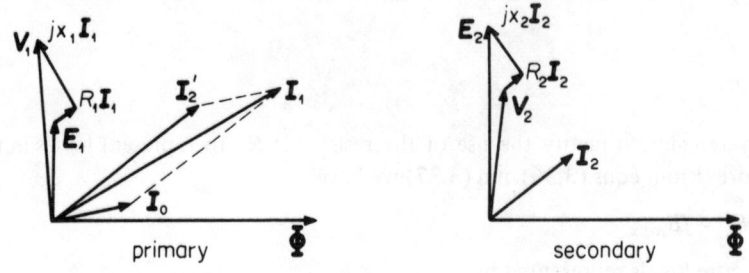

Figure 3.17 Phasor diagram for the transformer

$R_1 + jx_1$ and $R_2 + jx_2$ are zero. Instrument transformers, which are used to extend the voltage or current ranges of AC measuring instruments, are required to approximate as closely as possible to the ideal transformer; this is achieved by careful design, to minimise the unwanted terms in eqns (3.39) and (3.40).

Simplification of the equivalent circuit

Practical iron-cored transformers are usually designed so that under normal working conditions the volt drop in R_1 and x_1 is small in comparison with V_1, and I_0 is small in comparison with the load current I_1. The shunt components R_c and X_m may then be transferred to the input terminals with very little loss of accuracy (figure 3.18(a)). The secondary quantities R_2 and x_2 may be replaced by equivalent quantities on the primary side, using the impedance-transforming property of the ideal transformer, as shown in figure 3.18(b). The actual primary quantities R_1 and x_1 can then be combined with the referred secondary quantities R_2' and x_2' to give the equivalent primary quantities R_e and X_e (figure 3.18(c)). This is the normal form of the equivalent circuit for small power transformers; it simplifies the analysis because the current I_0 is determined solely by the applied primary voltage V_1, and the parameters are readily determined from simple tests. Figure 3.18 shows the elements referred to the primary side, and a 'mirror image' circuit is readily derived with all the elements referred to the secondary. This form is useful when the transformer is operated in an inverted

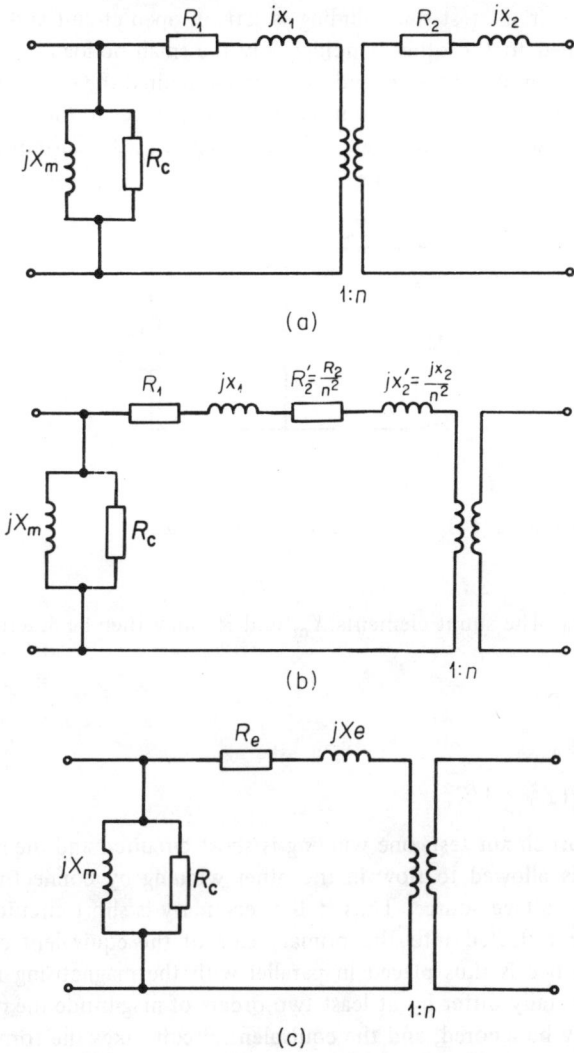

Figure 3.18 Simplified equivalent circuit of the transformer

mode, with a voltage applied to the secondary and a load connected to the primary.

Determination of the equivalent circuit parameters

The parameters of the approximate equivalent circuit are readily obtained from open-circuit and short-circuit tests, as follows.

In the open-circuit test one winding is left on open circuit and the normal voltage is applied to the other winding; only the small no-load current will be drawn from the supply. If the secondary is open-circuited the referred secondary current I_2' will be zero, and the equivalent circuit reduces to the form shown in figure 3.19. Measurements are made of the voltage V_0, the current I_0, and the

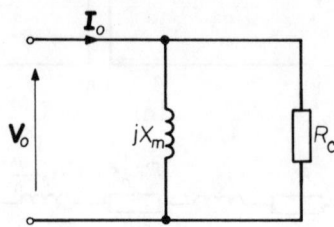

Figure 3.19 Equivalent circuit for the open–circuit test

input power P_0. The shunt elements X_m and R_c may then be determined from the relations

$$R_c = V_0^2/P_0 \tag{3.41}$$

$$Y_0 = I_0/V_0 \tag{3.42}$$

$$X_m = 1/\sqrt{(Y_0^2 - 1/R_c^2)} \tag{3.43}$$

For the short-circuit test, one winding is short circuited and the normal full-load current is allowed to flow in the other winding by connecting it to an adjustable low-voltage source. Thus if the secondary is short circuited, a short circuit will be reflected into the primary side of the equivalent circuit. The leakage impedance is thus placed in parallel with the magnetising impedance; since these generally differ by at least two orders of magnitude the magnetising impedance may be ignored, and the equivalent circuit takes the form shown in figure 3.20. Measurements are made of the voltage V_{sc}, the current I_{sc} and the input power P_{sc}; the series elements are then determined from the relations

$$R_e = R_{sc}/I_{sc}^2 \tag{3.44}$$

$$Z_e = V_{sc}/I_{sc} \tag{3.45}$$

$$X_e = \sqrt{(Z_e^2 - R_e^2)} \tag{3.46}$$

In practice the open-circuit measurements are usually made on the low-voltage side of the transformer and the short-circuit measurements on the high-voltage side (where the current will be lower). This is done merely for convenience, to avoid using higher voltages and currents than necessary; the same equations

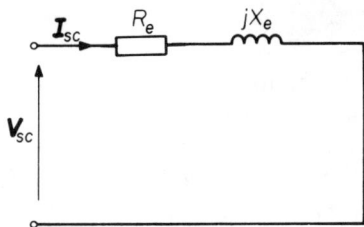

Figure 3.20 Equivalent circuit for the short-circuit test

apply, but for one test the element values will be referred to the primary and for the other they will be referred to the secondary. Conversion of secondary values to equivalent primary values (or vice versa) involves the turns ratio of the transformer; this is taken to be the primary–secondary voltage ratio measured in the open-circuit test.

Regulation and efficiency

Because of the volt drop in the series impedances of the windings, the transformer secondary terminal voltage will vary with the load current. The *voltage regulation* is defined as

$$\epsilon = \frac{\text{(no-load voltage)} - \text{(full-load voltage)}}{\text{(no-load voltage)}} \tag{3.47}$$

assuming a constant applied primary voltage.

The *efficiency* of the transformer is defined as

$$\eta = \frac{\text{output power}}{\text{input power}} \tag{3.48}$$

$$= \left\{ 1 - \frac{\text{losses}}{\text{output} + \text{losses}} \right\} \tag{3.49}$$

Power transformers have very high efficiencies — even a 1 kVA transformer will usually have an efficiency greater than 90 per cent — and the efficiency improves with increasing size. As with DC machines (section 2.3), the efficiency is calculated from eqn (3.49) by determining the losses, and not from eqn (3.48) by measuring the input and output power; the power measurements cannot be made with sufficient accuracy.

Auto-transformers

It is not essential for a transformer to have a separate secondary winding; the alternating core flux will induce an EMF in each turn of the primary winding, and the output can be taken from a portion of the same winding. This single-winding transformer is termed an *auto-transformer* [2]. Because part of the winding is effectively common to the primary and the secondary, there can be a substantial saving in weight and cost if the ratio of input to output voltage is less than about 3:1. Auto-transformers can be used only when electrical isolation is not required between the input and the output. An elegant application of the auto-transformer principle is the variable transformer, which has a single-layer coil wound on a toroidal core. The output is taken from a carbon brush which makes contact with the surface of the coil; the brush can be moved from one end of the coil to the other, thus varying the output voltage.

Problems

3.1. In a balanced three-phase four-wire system, show that no current will flow in the neutral conductor. Hence show that any balanced three-phase system (three-wire or four-wire) may be resolved into three separate single-phase systems with the voltages, currents and element values in the phases equal to the equivalent star values for the original system.

3.2. A three-phase load consists of three equal impedances of magnitude Z and phase angle ϕ. If the load is star connected, show that the instantaneous flow of power into the load is constant, with a value given by

$P = \sqrt{3} VI \cos \phi$

where V is the RMS line voltage, I is the RMS line current and $\cos \phi$ is the load power factor. Show that this expression also holds when the load is delta connected, and find the ratio of the line currents in the two cases.

3.3. The Scott connection of two transformers, shown in figure 3.21, is used for

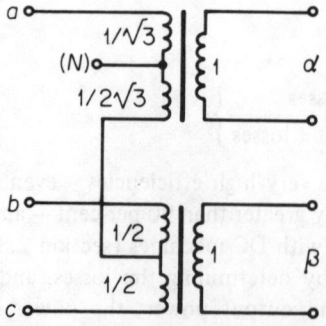

Figure 3.21 Scott transformer connection

obtaining a two-phase supply α, β from a three-phase supply a, b, c with an optional neutral connection N. The transformers (which may be assumed to be ideal) have tapped primary windings, with the turns ratios indicated in the diagram. Construct a voltage phasor diagram for the system, and verify that the system will transform voltages from three to two phases and vice versa.

3.4. The ideal gyrator is a circuit element invented by Tellegen to complete the set of linear passive elements (resistor, capacitor, inductor, transformer, gyrator). The circuit symbol for a gyrator is shown in figure 3.22, and the defining equations are

$$v_1 = ki_2$$

$$v_2 = ki_1$$

Compare the properties of the gyrator with those of an ideal transformer. If a capacitance C is connected to the output of the gyrator, what does the voltage–current relation at the input represent?

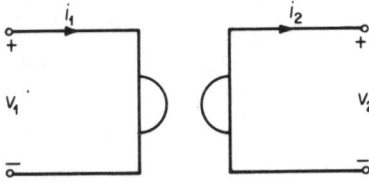

Figure 3.22 Ideal gyrator

3.5. A power transformer has the following nominal rating: primary 220 V, 50 Hz; secondary 660 V, 7.7 A, 5.1 kVA. Open-circuit and short-circuit tests gave the following results. With the secondary winding on open circuit, measurements taken at the *primary* were 220 V, 1.18 A, 65.5 W; the secondary voltage was 669 V. With the primary winding short circuited, measurements taken at the *secondary* were 13.2 V, 7.70 A, 69.5 W.

Calculate the parameters of the approximate equivalent circuit referred to the primary side of the transformer. Hence calculate (a) the secondary terminal voltage, (b) the efficiency, (c) the voltage regulation, when the transformer supplies its rated full-load secondary current of 7.7 A at unity power factor, with a primary supply voltage of 220 V.

3.6. In addition to the mains frequency of 50 Hz, a frequency of 400 Hz is commonly encountered in AC control systems. Consider two transformers of similar rating, one designed for operation at 50 Hz and the other for 400 Hz. When each transformer is operating at its rated voltage and frequency the peak core flux density is 1.3 T and the magnetising current is 5 per cent of the full-load primary current. The transformer cores are made from 4 per cent silicon steel, with a lamination thickness appropriate to the frequency.

By considering the core flux density, the magnetising current and the core loss, explain what will happen if (a) the 400 Hz transformer is operated at 50 Hz; (b) the 50 Hz transformer is operated at 400 Hz. In each case the normal rated voltage is applied to the primary of the transformer. Also explain why the 400 Hz transformer will be smaller and lighter than the 50 Hz transformer.

References

1 H. Waddicor, *Principles of Electric Power Transmission*, 5th ed. (London: Chapman and Hall, 1964).
2 M. G. Say, *Alternating Current Machines*, 5th ed. (London: Pitman, 1983).

4   Introduction to AC Machines

4.1 Introduction

The commutator in a DC machine performs a complex function which we have not attempted to analyse in detail; the performance equations, on the other hand, are relatively simple. With the AC synchronous and induction machines, the situation is reversed. The absence of a commutator simplifies both the structure and the detailed analysis, but the machine equations are more complex and the basic theory is conceptually more difficult. In this chapter we develop some of the principles which are common to all AC machines by considering the magnetic field set up by currents flowing in the windings. It is convenient to start with the AC generator introduced in the previous chapter.

The simple model of a generator introduced in section 3.2 has a rotor with prominent poles (a 'salient pole' rotor). Many practical machines are in fact made with a rotor structure of this kind, but the general theory is quite complex and is best handled by the methods introduced in chapter 7. A simpler theory results if the rotor is cylindrical, giving a uniform airgap (figure 4.1), and this form of construction is used for mechanical reasons in the high-speed turbine-driven generators of modern power stations. We have seen that a sinusoidal generated voltage is desirable, and this implies a sinusoidal variation of the flux density round the rotor when there is a single stator armature coil spanning a

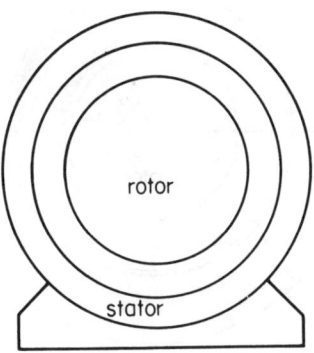

Figure 4.1   AC machine with cylindrical rotor

diameter. There are methods of arranging the coils in a practical armature winding to eliminate or reduce some of the harmonics when the flux-density variation is not a pure sinusoid, but these are design details that do not concern us here. Whatever may be achieved by subtle design of the armature winding, it is generally necessary to make the flux density variation as nearly sinusoidal as possible. With a salient pole rotor this is accomplished by shaping the poles; with a cylindrical rotor it can only be done by distributing the conductors of the field winding in a particular way.

4.2 Distributed windings and the airgap magnetic field

The coils forming the stator and rotor windings of a practical machine are usually arranged so that the conductors are distributed round the stator and rotor, instead of being concentrated at a number of points as they are in the simple models so far considered. To understand the operation of the machine it is necessary to calculate the magnetic field in the airgap from a knowledge of the current in the winding and the way in which the conductors are distributed; this may be done by an extension of the magnetic circuit concept.

MMF of a distributed winding

Figure 4.2 shows a cross-section through a cylindrical-rotor machine, with conductors distributed round the stator and rotor surfaces. Application of Ampère's

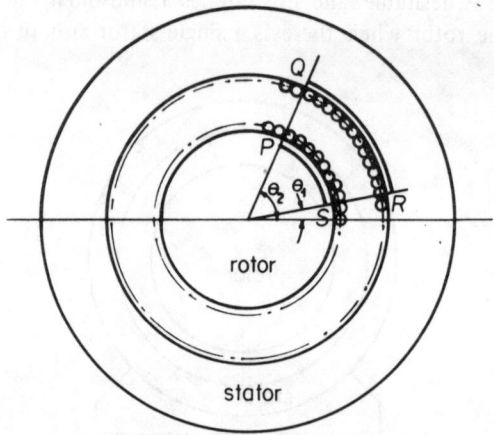

Figure 4.2   Distributed stator and rotor windings

circuital law to the path PQRS gives

$$\Sigma i = \oint H \cdot ds = \int_P^Q H \cdot ds + \int_Q^R H \cdot ds + \int_R^S H \cdot ds + \int_S^P H \cdot ds \quad (4.1)$$

In this equation, Σi is the total current carried by the conductors within the boundary PQRS. If the permeability of the stator and rotor iron is high, the magnetic potential drops along the iron paths QR and SP will be negligible in comparison with the airgap potential drops along PQ and RS. It is reasonable to assume that H is nearly uniform along a radial path in the airgap, so eqn (4.1) becomes

$$\Sigma i = \int_P^Q H \cdot ds + \int_R^S H \cdot ds$$

$$= gH_r(\theta_2) - gH_r(\theta_1) \quad (4.2)$$

where g is the radial length of the airgap, and H_r is the radial component of the magnetising force. Usually, the magnetic field is nearly radial, so that $H_r \approx |H|$; we shall therefore drop the subscript r, with the understanding that the quantity H appearing in the equation is, strictly speaking, the radial component. At some point in the airgap, the value of H will be zero; let the corresponding value of θ be θ_0. At any other value of θ we have

$$\Sigma i = gH(\theta) - gH(\theta_0)$$

$$= gH(\theta) \quad (4.3)$$

where Σi is now the sum of the currents in the conductors between θ_0 and θ. This quantity Σi is known as the MMF of a winding; since it depends on θ, it may be denoted by $F(\theta)$. Thus the radial component of the magnetic field depends on the airgap length g and also on the quantity F, which specifies the way in which the winding conductors are distributed.

Linear current density

Since the magnetic field is produced by a distribution of current-carrying conductors, it is convenient to introduce a quantity which describes the way in which the current is distributed. This quantity is the linear current density K; it is a vector defined as follows. Let ds be an element of length along the iron surface of the stator or the rotor, measured in the circumferential direction. If di represents the amount of current in the conductors in this length ds, then

$$di = K\,ds \quad (4.4)$$

The magnitude of K is thus the current per unit length along the surface, measured perpendicular to the current flow, and the direction of K is the direction of the current. This is similar to the definition of the ordinary current density J (section 1.2), which is also a vector in the direction of current flow but with a magnitude equal to the current per unit area.

Let K_1 and K_2 be the linear current densities on the stator and rotor respectively. The MMF is given by

$$F = \Sigma i = \int di_1 + \int di_2 = \int K_1 \, ds_1 + \int K_2 \, ds_2 \tag{4.5}$$

If r_1 and r_2 are the radii of the stator and rotor surfaces respectively, then

$$F(\theta) = \Sigma i = \int_{\theta_0}^{\theta} K_1 r_1 \, d\theta + \int_{\theta_0}^{\theta} K_2 r_2 \, d\theta \tag{4.6}$$

This expression for the MMF may be substituted in eqn (4.3) to give a direct relationship between the flux density and the current density

$$B(\theta) = \mu_0 H(\theta) = \frac{\mu_0 r_1}{g} \int_{\theta_0}^{\theta} K_1 \, d\theta + \frac{\mu_0 r_2}{g} \int_{\theta_0}^{\theta} K_2 \, d\theta \tag{4.7}$$

Sinusoidally distributed windings

If there is current in only one of the two windings, then eqn (4.7) becomes

$$B(\theta) = \frac{\mu_0 r}{g} \int_{\theta_0}^{\theta} K \, d\theta \tag{4.8}$$

where r and K represent either r_1 and K_1 or r_2 and K_2. If the flux density is to be cosinusoidal, then we may put

$$B(\theta) = B_m \cos \theta \tag{4.9}$$

From eqn (4.8) the required current density is

$$K(\theta) = \frac{g}{\mu_0 r} \frac{dB}{d\theta}$$

$$= -\frac{g B_m}{\mu_0 r} \sin \theta$$

$$= -K_m \sin \theta \tag{4.10}$$

where

$$B_m = \frac{\mu_0 r}{g} K_m \tag{4.11}$$

We thus require a sinusoidal current density in order to produce a cosinusoidal flux density: this is illustrated in figure 4.3, where the size of the small circles

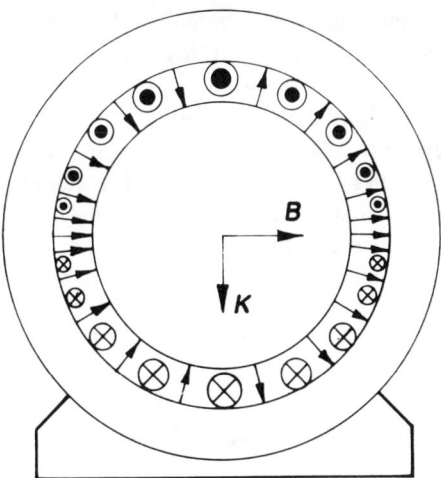

Figure 4.3 Sinusoidal current density and cosinusoidal flux density

represents the magnitude of the current density. Note the right-angle relationship between the current-density and flux-density distributions in figure 4.5. This relationship is illustrated by the vectors K and B in figure 4.5, which point in the directions of maximum K and B respectively. These vectors are formally defined in section 4.3.

A sinusoidal current density can, in principle, be produced by distributing the conductors in a non-uniform way. If each conductor carries a current of magnitude i, then the conductor density — the number of conductors per unit length — must vary in the same way as the magnitude of the current density. The small circles in figure 4.3 thus represent the conductor density as well as the current density, and the winding is said to be sinusoidally distributed. Formally we may put

$$K = -K_m \sin \theta = -Zi \sin \theta \tag{4.12}$$

where Z is the maximum number of conductors per unit length, and eqn (4.11) may be written

$$B_m = \frac{\mu_0 r}{g} Zi \tag{4.13}$$

In some special machines such as synchros, which are used as rotary position control devices [1, 2] a close approximation to a true sinusoidal distribution is used. But this is uneconomic for normal AC motors and generators, and for these machines a simpler form of winding is used. It will be shown in section 4.8 that the simple winding can give an acceptably close approximation to the ideal sinusoid.

The concepts of a sinusoidally distributed winding and a sinusoidal magnetic field are of fundamental importance in AC machines. The requirement of a rotor winding of this kind arose from the need to generate sinusoidal voltages in the stator coils. If these coils form part of a distributed stator winding, the magnetic field produced by currents in this winding should have the same form as the rotor field; and this implies that the stator winding should also be sinusoidally distributed. Optimum machine performance is obtained with sinusoidal quantities, and by a happy coincidence this also gives the simplest mathematical treatment. An exactly sinusoidal distribution will therefore be assumed in developing the theory of AC machines.

4.3 Combination of sinusoidally distributed fields

Consider a winding carrying a current i_1, with a conductor distribution such that the current density is

$$K_1 = -K_{1m} \sin(\theta - \alpha)$$
$$= -Z_1 i_1 \sin(\theta - \alpha) \tag{4.14}$$

The magnetic flux density produced by this winding will be

$$B_1 = B_{1m} \cos(\theta - \alpha) \tag{4.15}$$

where

$$B_{1m} = \frac{\mu_0 r_1}{g} Z_1 i_1 \tag{4.16}$$

This sinusoidally distributed field pattern is illustrated in figure 4.4.

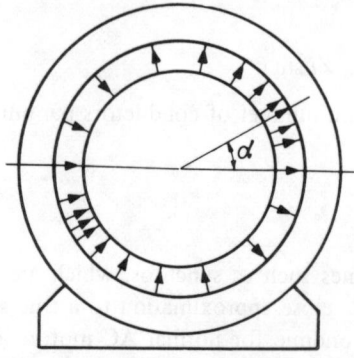

Figure 4.4 Sinusoidal field B_1

INTRODUCTION TO AC MACHINES

Suppose that there is a second sinusoidally distributed winding carrying a current i_2, with a conductor distribution such that the current density is

$$K_2 = -K_{2m} \sin(\theta - \beta)$$
$$= -Z_2 i_2 \sin(\theta - \beta) \tag{4.17}$$

This will give rise to a second magnetic field

$$B_2 = B_{2m} \cos(\theta - \beta) \tag{4.18}$$

where

$$B_{2m} = \frac{\mu_0 r_2}{g} Z_2 i_2 \tag{4.19}$$

This is illustrated in figure 4.5. The total magnetic field B is the sum of the separate fields

$$B = B_1 + B_2$$
$$= B_{1m} \cos(\theta - \alpha) + B_{2m} \cos(\theta - \beta) \tag{4.20}$$

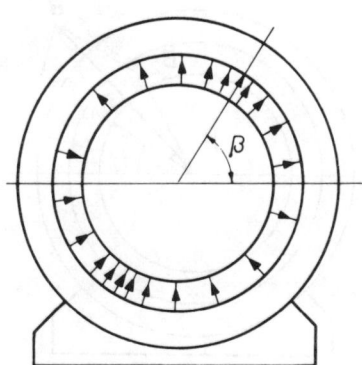

Figure 4.5   Sinusoidal field B_2

Since the sum of two sinusoids is another sinusoid, this may be written as

$$B = B_m \cos(\theta - \gamma) \tag{4.21}$$

and the magnetic field distribution is shown in figure 4.6.

Vector representation of sinusoidal fields

In AC circuit theory the manipulation of sinusoids is simplified by the use of rotating vectors or phasors. A similar device can be used with sinusoidally distri-

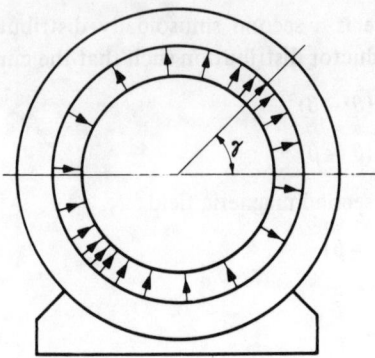

Figure 4.6 Sinusoidal field $B = B_1 + B_2$

buted magnetic fields. We represent a field of maximum value B_m by a radius vector of length B_m, pointing in the direction of the maximum field intensity. Thus in figure 4.7, the field B_1 is represented by a vector $\boldsymbol{B}_1$ of length B_{1m},

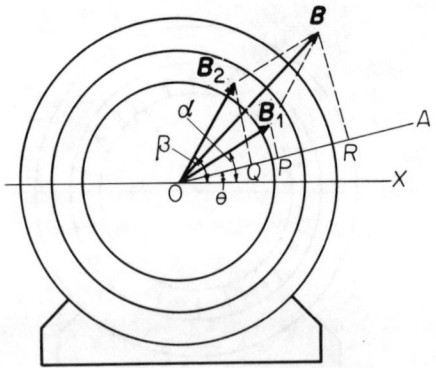

Figure 4.7 Space phasor representation of sinusoidal fields

making an angle α with the reference axis OX; and the field B_2 is represented by a vector $\boldsymbol{B}_2$ of length B_{2m}, making an angle β with the reference axis. Let OA be a line making an angle θ with OX. The projection of $\boldsymbol{B}_1$ on OA is

$$\text{OP} = B_{1m} \cos(\alpha - \theta) = B_{1m} \cos(\theta - \alpha) \tag{4.22}$$

and the projection of $\boldsymbol{B}_2$ on OA is

$$\text{OQ} = B_{2m} \cos(\beta - \theta) = B_{2m} \cos(\theta - \beta) \tag{4.23}$$

Thus the value of the magnetic field at any angle θ in the airgap is equal to the projection of the respective vector on to a line making an angle θ with the reference axis. Now consider the vector sum $B_1 + B_2$. The projection of this vector on to OA is given by

$$OR = OP + PR$$
$$= OP + OQ$$
$$= B_{1m} \cos(\theta - \alpha) + B_{2m} \cos(\theta - \beta) \qquad (4.24)$$

and it therefore represents the total magnetic field B. The length of this vector is B_m, and it makes an angle γ with OX.

This procedure is exactly analogous to the phasor addition of sinusoidally time-varying voltages or currents. Following Chapman [2], we shall use the term 'space phasor' for the vector representing the sinusoidally distributed field, to avoid confusion with the magnetic field vector. Magnetic flux density is a vector quantity possessing magnitude and direction. In developing the theory of electrical machines we are mainly concerned with the radial component of the flux density vector, and in particular with the variation of the scalar magnitude of this component round the airgap. It is the spatial variation of a scalar magnitude which is described by the 'vector' introduced in this section.

Sinusoidally distributed current density may likewise be represented by a space phasor. Thus in figure 4.3, B is the space phasor representing the flux-density distribution, and K is the space phasor representing the current-density distribution.

4.4 Torque from sinusoidally distributed windings

Suppose that the rotor and stator each have sinusoidally distributed windings. If the stator winding carries a current i_1, it will produce a magnetic field of the form

$$B_1 = B_{1m} \cos(\theta - \alpha) \qquad [4.15]$$

and the stator will behave like a permanent magnet with north and south poles. Likewise, if the rotor carries a current i_2 it will produce a magnetic field of the form

$$B_2 = B_{2m} \cos(\theta - \beta) \qquad [4.18]$$

and the rotor will also behave like a permanent magnet. The magnetic axes of the stator and rotor will be displaced by an angle $\delta_{12} = \alpha - \beta$, and there will be a torque on the rotor tending to pull its poles into alignment with the stator poles. This is illustrated in figure 4.8, where the bending of the lines of force is greatly exaggerated. The magnetic field cannot be purely radial, or there would be no torque on the rotor. This follows from the tangential Maxwell stress formula

$$t_s = \frac{B_n B_s}{\mu_0} \qquad [1.43]$$

If the field were purely radial, we would have $B_s = 0$, and therefore $t_s = 0$. It is shown in appendix A that a non-uniform radial magnetic field must be accompanied by a circumferential component; when the stator and rotor axes are displaced, the radial and circumferential components combine to give a field pattern of the form shown in figure 4.8.

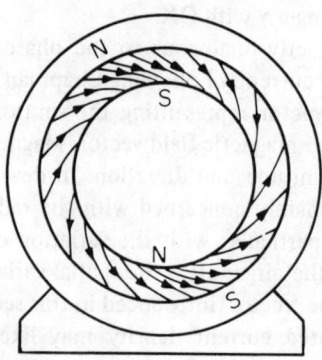

Figure 4.8 Magnetic field for displaced stator and rotor axes

The torque tending to align the magnetic axes of the stator and rotor may be calculated in a number of ways. Energy methods are commonly used [3, 4], but a more direct method is to consider the force acting on the current in each element of the rotor surface; this is merely an extension of the method already used for the DC machine, and it is shown in appendix A that this is equivalent to evaluating the Maxwell stress. Let di_2 be the current in an element of arc $d\theta$; this current will interact with the magnetic field B_1 of the stator to give a force

$$d\boldsymbol{F} = \boldsymbol{l} \times \boldsymbol{B}_1 \, di_2 \qquad (4.25)$$

where l is the axial length of the conductors in the arc $d\theta$. Only the radial component of $\boldsymbol{B}_1$ will give rise to a circumferential component of $d\boldsymbol{F}$, and the contribution to the torque is thus

$$dT = -r_2 l B_{1r} \, di_2 \qquad (4.26)$$

where B_{1r} is the radial component of $\boldsymbol{B}_1$ and r_2 is the radius of the rotor. The negative sign arises from the convention that counter-clockwise torque is positive. Since the value of $\boldsymbol{B}_1$ calculated from the current distribution is in fact the radial component (see section 4.2), eqn (4.15) may be written in the form

$$B_{1r} = B_{1m} \cos(\theta - \alpha) \qquad (4.27)$$

The current di_2 is given by

$$di_2 = K_2 r_2 \, d\theta$$
$$= -Z_2 i_2 r_2 \sin(\theta - \beta) \, d\theta \qquad (4.28)$$

Since the rotor current i_2 is related to the maximum value B_{2m} of the rotor magnetic field by eqn (4.19), the last equation may be written as

$$di_2 = -\frac{g}{\mu_0} B_{2m} \sin(\theta - \beta) \, d\theta \qquad (4.29)$$

Substitution of these expressions for B_{1r} and di_2 into eqn (4.26) and integrating to obtain the total torque gives

$$T = \int_0^{2\pi} \frac{r_2 lg}{\mu_0} B_{1m} \cos(\theta - \alpha) B_{2m} \sin(\theta - \beta) \, d\theta$$

$$= k B_{1m} B_{2m} \sin \delta_{12} \quad \text{newton metres} \qquad (4.30)$$

where

$$k = \frac{\pi r_2 lg}{\mu_0} \qquad (4.31)$$

and

$$\delta_{12} = \alpha - \beta$$

Equation (4.30) is an important result, which shows that the alignment torque T varies as the sine of the angular separation δ_{12} between the stator and rotor magnetic axes.

The relation between the component fields B_1 and B_2 and the total field B is given by the space phasor diagram of figure 4.9. Application of the sine rule to this diagram gives

$$\frac{B_{1m}}{\sin \delta_2} = \frac{B_{2m}}{\sin \delta_1} = \frac{B_m}{\sin \delta_{12}} \qquad (4.32)$$

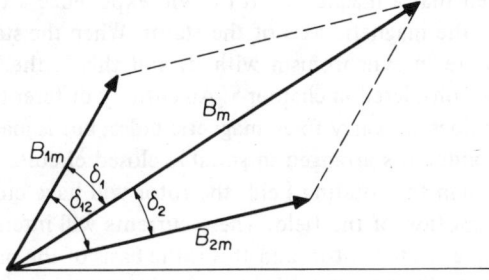

Figure 4.9 Space phasor diagram for displaced stator and rotor axes

The torque equation may therefore be written as

$$T = kB_{1m}B_{2m} \sin \delta_{12} = kB_{1m}B_m \sin \delta_1 = kB_{2m}B_m \sin \delta_2 \qquad (4.33)$$

showing that the torque is proportional to the product of any two of the field components and the sine of the angle between them.

4.5 The rotating magnetic field

The equation

$$B = B_m \cos(\theta - \psi) \qquad (4.34)$$

represents a sinusoidally distributed magnetic field with its axis inclined at an angle ψ. Suppose that this magnetic field is produced by a distributed winding on the stator, and that by some means the angle ψ is made to increase uniformly with time, so that

$$\psi = \psi_0 + \omega t \qquad (4.35)$$

where ω is a constant. For convenience, let $\psi_0 = 0$. The axis, and thus the sinusoidal magnetic field pattern, is rotating in the positive (counter-clockwise) direction with an angular velocity ω. The expression for the magnetic field becomes

$$\begin{aligned} B &= B_m \cos(\theta - \omega t) \\ &= B_m \cos(\omega t - \theta) \end{aligned} \qquad (4.36)$$

and this is the equation of a rotating magnetic field. At any particular instant of time t, eqn (4.36) shows that the field is sinusoidally distributed round the airgap, with its axis inclined at an angle ωt. The equation also shows that at any particular angle θ the field varies sinusoidally with time, but with a phase lag of θ. This is exactly analogous to a sinusoidal travelling wave of the form $\cos(x - ut)$, encountered in transmission line theory.

It has been seen that a magnetised rotor will experience a torque tending to align its axis with the magnetic axis of the stator. When the stator field rotates, the rotor will rotate in synchronism with it, and this is the basis of the synchronous machine considered in chapter 5. An entirely different kind of machine results if the rotor does not carry fixed magnetic poles, but is made of conducting material or has conductors arranged in suitable closed circuits. If the rotor runs at a lower speed than the rotating field, the rotor will have currents induced in it by the relative motion of the field. These currents will interact with the field to produce a torque on the rotor, and this is the basis of the induction machine studied in chapter 6. We now consider how a rotating stator magnetic field may be produced from fixed windings.

Production of a rotating magnetic field

Suppose that the stator is provided with two sinusoidally distributed windings α and β, which are similar in all respects except that their axes are at $\theta = 0$ and $\theta = \pi/2$ respectively. This arrangement is termed a *two-phase winding*. If the windings carry currents i_α and i_β, the flux density produced by each winding will be

$$\left. \begin{array}{l} B_\alpha = B_{\alpha m} \cos \theta = c i_\alpha \cos \theta \\ B_\beta = B_{\beta m} \cos(\theta - \pi/2) = c i_\beta \cos(\theta - \pi/2) \end{array} \right\} \quad (4.37)$$

where

$$c = \frac{\mu_0 r_1}{g} Z_1 \quad (4.38)$$

The total flux density is

$$B = B_\alpha + B_\beta = B_m \cos(\theta - \psi) \quad (4.39)$$

The amplitude B_m and space phase ψ of the resultant field are given by the space phasor diagram of figure 4.10, from which

$$\left. \begin{array}{l} B_{\alpha m} = B_m \cos \psi \\ B_{\beta m} = B_m \sin \psi \end{array} \right\} \quad (4.40)$$

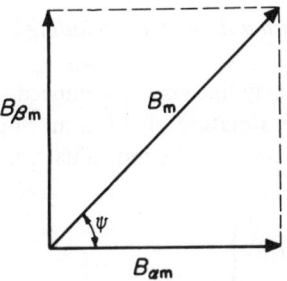

Figure 4.10 Space phasor diagram for a two-phase winding

For a pure rotating field we require B_m = constant and $\psi = \omega t$; from eqns (4.40) and (4.37) we thus have

$$\left. \begin{array}{l} c i_\alpha = B_{\alpha m} = B_m \cos \omega t \\ c i_\beta = B_{\beta m} = B_m \sin \omega t \end{array} \right\} \quad (4.41)$$

Equation (4.41) will be true if the currents are given by

$$\left.\begin{aligned} i_\alpha &= I_m \cos \omega t \\ i_\beta &= I_m \sin \omega t = I_m \cos(\omega t - \pi/2) \end{aligned}\right\} \quad (4.42)$$

These are the equations of two-phase alternating currents, and it can be concluded that a rotating magnetic field will be produced by two-phase currents flowing in a two-phase winding. This result may also be obtained by direct substitution of eqns (4.42) in eqns (4.37); then

$$\begin{aligned} B &= cI_m(\cos \omega t \cos \theta + \sin \omega t \sin \theta) \\ &= B_m \cos(\omega t - \theta) \end{aligned} \quad (4.43)$$

where

$$B_m = cI_m = \frac{\mu_0 r_1}{g} Z_1 I_m \quad (4.44)$$

For simplicity, the two-phase winding has been chosen with the β phase leading the α phase by 90°. It is customary to show the β phase lagging by 90°, and this is the convention adopted in chapter 7. This makes no essential difference to the analysis; with the currents given by eqn (4.42), the field would be $B = B_m \cos(\omega t + \theta)$, signifying rotation in the negative direction; reversing the sign of i_β (changing the phase from 90° lag to 90° lead) will restore the positive direction of rotation.

Rotating magnetic field with a three-phase winding

A rotating magnetic field may likewise be produced from a symmetrical m-phase supply, provided that the armature also has an m-phase symmetrical winding. Thus for three phases, the fields of the individual armature phases will be

$$\left.\begin{aligned} B_a &= ci_a \cos \theta \\ B_b &= ci_b \cos(\theta - 2\pi/3) \\ B_c &= ci_c \cos(\theta - 4\pi/3) \end{aligned}\right\} \quad (4.45)$$

If the windings are supplied from a three-phase source, the currents will be

$$\left.\begin{aligned} i_a &= I_m \cos \omega t \\ i_b &= I_m \cos(\omega t - 2\pi/3) \\ i_c &= I_m \cos(\omega t - 4\pi/3) \end{aligned}\right\} \quad (4.46)$$

The total field is $B = B_a + B_b + B_c$, and the airgap flux density is therefore

$$\begin{aligned} B = cI_m\{&\cos \omega t \cos \theta + \cos(\omega t - 2\pi/3)\cos(\theta - 2\pi/3) + \\ &+ \cos(\omega t - 4\pi/3)\cos(\theta - 4\pi/3)\} \end{aligned}$$

$$= \tfrac{1}{2}cI_m\{\cos(\omega t - \theta) + \cos(\omega t + \theta) + \cos(\omega t - \theta) + $$
$$+ \cos(\omega t + \theta - 4\pi/3) + \cos(\omega t - \theta) + \cos(\omega t + \theta - 8\pi/3)\}$$
$$= B_m \cos(\omega t - \theta) \tag{4.47}$$

where

$$B_m = \frac{3c}{2} I_m = \frac{3\mu_0 r_1}{2g} Z_1 I_m \tag{4.48}$$

Reversal of the direction of rotation

Interchanging i_b and i_c would give

$$\left. \begin{array}{l} i_{a'} = i_a = I_m \cos \omega t \\ i_{b'} = i_c = I_m \cos(\omega t - 4\pi/3) = I_m \cos(\omega t + 2\pi/3) \\ i_{c'} = i_b = I_m \cos(\omega t - 2\pi/3) = I_m \cos(\omega t + 4\pi/3) \end{array} \right\} \tag{4.49}$$

With these expressions for current, we obtain

$$B = B_m \cos(\omega t + \theta) \tag{4.50}$$

which represents a magnetic field rotating in the opposite direction. Thus the direction of rotation of the field may be reversed just by reversing the phase sequence of the supply.

4.6 Voltage induced by a rotating magnetic field

Alternating currents flowing in suitable windings will produce a rotating magnetic field; we now calculate the voltage induced by the rotating field, and find a relationship between the voltage and current for each phase of the winding under balanced operating conditions.

A two-phase machine winding is a natural choice, and this form of winding was used in the early days of AC systems. Since modern power systems use three phases for economic reasons, most industrial machines have three-phase windings. Domestic induction motors are usually single-phase machines, but these are special cases which are considered later in chapter 6. The two-phase machine is now confined almost entirely to AC servo systems [1]; but the principles are virtually the same no matter how many phases are used to produce a uniform rotating field, and only the simpler two-phase model will be analysed in detail. The results may often be generalised to other numbers of phases by inspection, and it is shown in appendix B that there is a mathematical transformation which enables the performance of a three-phase machine to be calculated from the analysis of an equivalent two-phase machine. This transformation holds for all conditions of operation: balanced or unbalanced, steady-state or transient; so there is no loss of generality in using a two-phase model.

Voltage induced in a winding

Consider a single-turn coil in the airgap, as shown in figure 4.11, and a rotating magnetic field given by

$$B = B_m \cos(\omega t - \theta) \qquad [4.43]$$

The flux linking this coil is given by

$$\Phi = \int_{-\pi/2}^{\pi/2} Blr \, d\theta \qquad (4.51)$$

where r is the radial distance of a coil side from the axis, and l is the length of each coil side. Evaluating the integral in eqn (4.51) gives

$$\Phi = \int_{-\pi/2}^{\pi/2} B_m lr \cos(\omega t - \theta) \, d\theta$$

$$= 2rlB_m \sin(\omega t + \pi/2) \qquad (4.52)$$

and the induced EMF in the coil is thus

$$e = \frac{d\Phi}{dt} = 2rl\omega B_m \cos(\omega t + \pi/2) \qquad (4.53)$$

This result could also be obtained from the flux cutting rule (eqn 1.5); the velocity of the field relative to the conductor is $u = \omega r$, and the magnitude of the EMF induced in each conductor is

$$e_c = Blu = rl\omega B_m \cos(\omega t + \pi/2) \qquad (4.54)$$

Equation (4.54) is an important step in developing the theory of induction machines (section 6.3), but the use of the flux cutting rule may seem dubious

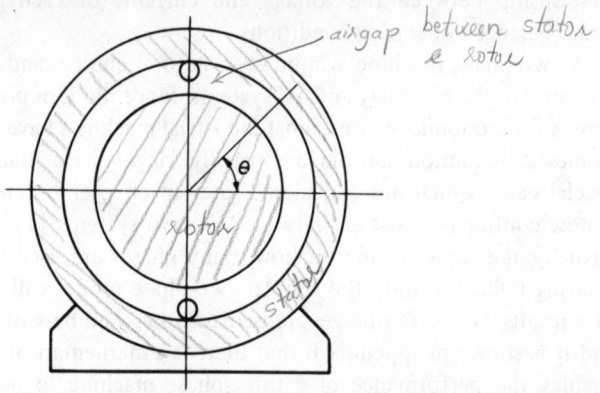

Figure 4.11 Single-turn stator coil

INTRODUCTION TO AC MACHINES

in view of the warning given in section 1.3; the velocity u is certainly not the velocity relative to any material part. The derivation of eqn (4.53), however, is not open to dispute; this shows that the EMF in each conductor is given by eqn (4.54), and it provides the formal justification for using the flux cutting rule with rotating or travelling fields.

If the single-turn coil is replaced by a sinusoidally distributed winding, the induced EMF may be found by integration; for the α-phase winding the result is

$$e_\alpha = \pi r^2 l Z \omega B_m \cos(\omega t + \pi/2) \tag{4.55}$$

Relationship between phase voltage and current

Equation (4.55) gives the EMF induced in the α phase by the rotating magnetic field. When the field is itself produced by currents flowing in the two phases of the winding, there is an important relationship between the phase voltage and current. If the winding resistance and leakage reactance may be neglected, the terminal voltage v_α is equal to the induced EMF e_α. The maximum flux density is related to the maximum current in the winding by the equation

$$B_m = \frac{\mu_0 r}{g} Z I_m \tag{4.44}$$

and the terminal voltage is therefore given by

$$v_\alpha = \frac{\pi r^3 l Z^2 \mu_0}{g} \omega I_m \cos(\omega t + \pi/2) \tag{4.56}$$

The current flowing in this phase is

$$i_\alpha = I_m \cos \omega t \tag{4.42}$$

Comparison of the last two equations shows that (a) the voltage leads the current by $\pi/2$ radians; (b) the amplitude of the voltage is proportional to the amplitude of the current; (c) the amplitude of the voltage is proportional to the angular frequency ω. In terms of phasors we have

$$V_\alpha = j\omega \left[\frac{\pi r^3 l Z^2 \mu_0}{g} \right] I_\alpha \tag{4.57}$$

showing that this phase of the machine behaves as an inductance of value

$$L = \frac{\pi r^3 l Z^2 \mu_0}{g} \text{ henrys} \tag{4.58}$$

It is readily shown that the terminal voltage of the β phase is given by

$$v_\beta = \frac{\pi r^3 l Z^2 \mu_0}{g} \omega I_m \sin(\omega t + \pi/2) \tag{4.59}$$

while the current is

$$I_\beta = I_m \sin \omega t \tag{4.60}$$

Thus the β phase also behaves as an inductance of value L; the voltage and current have the same magnitudes as those in the α phase, but lag by an angle of $\pi/2$ radians. Similar results hold for a symmetrical m-phase machine; the currents and voltages have the same magnitudes in each phase, but there is a phase shift of $2\pi/m$ between adjacent phases.

Relationship between space and time phasors

Suppose that a machine has windings on the stator which produce a rotating field of the form

$$B_1 = B_{1m} \cos(\theta - \omega t - \alpha) \tag{4.61}$$

and windings on the rotor which produce a rotating field of the form

$$B_2 = B_{2m} \cos(\theta - \omega t - \beta) \tag{4.62}$$

These fields will combine to give a total field

$$B = B_m \cos(\theta - \omega t - \gamma) \tag{4.63}$$

and at any instant of time the magnetic field components may be represented by the space phasors B_1, B_2 and B. Figure 4.12 shows the space phasor diagram for

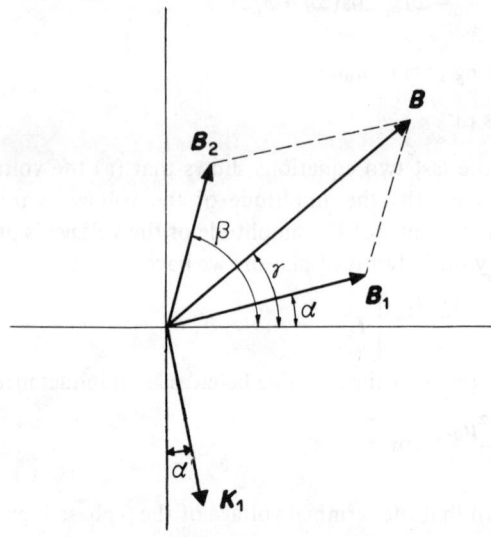

Figure 4.12 Space phasor diagram for stator and rotor fields

INTRODUCTION TO AC MACHINES

the instant $t = 0$, and the phasors may be imagined to rotate with angular velocity ω in the counter-clockwise direction. Each field component acting alone would induce EMFs in the phases of the windings. Consider the EMF e_1 induced in the α phase of the stator winding by the field component B_1. The previous analysis may be used if we let

$$\omega t' = \omega t + \alpha \tag{4.64}$$

Equation (4.61) then becomes

$$B_1 = B_{1m} \cos(\omega t' - \theta) \tag{4.65}$$

and eqn (4.55) gives the induced EMF

$$e_1 = CB_{1m} \cos(\omega t' + \pi/2) = CB_{1m} \cos(\omega t + \alpha + \pi/2) \tag{4.66}$$

where

$$C = \pi r^2 lZ\omega \tag{4.67}$$

Similarly the EMF e_2 induced by the field component B_2 will be

$$e_2 = CB_{2m} \cos(\omega t + \beta + \pi/2) \tag{4.68}$$

and the total EMF e induced by the total field B will be

$$e = CB_m \cos(\omega t + \gamma + \pi/2) \tag{4.69}$$

These EMFs may be represented by the time phasors E_1, E_2 and E shown in figure 4.13, which is drawn for the instant $t = 0$.

Currents must flow in the stator winding to produce the field B_1, and the phasor I_1 representing the current in the α phase will lag the EMF E_1 by 90° as shown in figure 4.13. But we have also seen that the sinusoidal current-density pattern is displaced by 90° from the flux-density pattern which it produces. Consequently the stator current density K_1 can be represented by the space phasor K_1, which lags B_1 by 90°, as shown in figure 4.12.

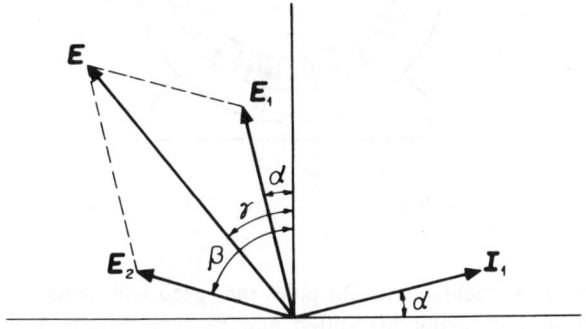

Figure 4.13 Time phasor diagram for one stator phase

There is an exact parallel between the space phasor diagram of figure 4.12 and the time phasor diagram of figure 4.13. The lengths of the time phasors are proportional to the lengths of the corresponding space phasors, and the angles between the phasors are the same in the two diagrams. The relationship between the field components represented by figure 4.12 and the phase voltage and current represented by figure 4.13 is crucial to the theory of AC machines; it leads directly to the equivalent circuit of the synchronous machine in chapter 5, and in a more subtle way to the equivalent circuit of the induction machine in chapter 6. The shift of 90° between the two diagrams will be suppressed in chapters 5 and 6, and time phasors will be drawn parallel to the corresponding space phasors.

4.7 Multi-pole fields

The machine windings so far considered produce two effective magnetic poles, and the field makes one revolution during one cycle of the AC supply. The length of arc between one pole and the next is known as a *pole pitch*, and we may therefore say that the field moves through two pole pitches during one cycle of the supply.

Suppose that we construct a machine having a four-pole rotor and a stator winding that produces a four-pole field (figure 4.14). During one cycle of the supply the field will move through two pole pitches as before, but this represents only half a revolution. The speed of the rotating field is thus $\omega/2$ for a four-pole

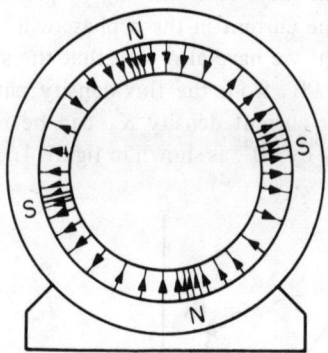

Figure 4.14 Four-pole field

machine, and for a machine with $2p$ poles the speed will be ω/p. The magnetic field will still be sinusoidally distributed, and it will go through a complete cycle in two pole pitches regardless of the number of poles on the machine. It is convenient, therefore, to work in terms of an electrical angle θ_e which increases by

2π for each complete cycle of the field. If the actual angular position or 'mechanical angle' is θ, then the electrical angle is given by

$$\theta_e = p\theta \tag{4.70}$$

This procedure may be justified formally as follows. The field components of a $2p$-pole two-phase winding will be given by

$$\left.\begin{array}{l} B_\alpha = ci_\alpha \cos p\theta \\ B_\beta = ci_\beta \cos(p\theta - \pi/2) \end{array}\right\} \tag{4.71}$$

If a two-phase supply is connected to the winding, the currents will be

$$\left.\begin{array}{l} i_\alpha = I_m \cos \omega t \\ i_\beta = I_m \cos(\omega t - \pi/2) \end{array}\right\} \tag{4.72}$$

The total field is $B = B_\alpha + B_\beta$; from eqns (4.71) and (4.72) this is

$$\begin{aligned} B &= cI_m \cos(\omega t - p\theta) \\ &= cI_m \cos p\left(\frac{\omega}{p}t - \theta\right) \end{aligned} \tag{4.73}$$

When the supply angular frequency is ω, eqn (4.73) shows that a $2p$-pole field will rotate with angular velocity ω/p. This quantity is termed the *synchronous angular velocity*, denoted by ω_s and measured in radians per second.

Analysis of multi-pole machines

If $p\theta$ is replaced by θ_e, eqns (4.71) and (4.73) take the same form as the equations for a two-pole winding, and the analysis of two-pole machines may be applied directly to multi-pole machines. For the rest of this book a two-pole machine will be assumed unless the contrary is stated. To extend the analysis to multi-pole machines, it is only necessary to multiply the synchronous speed by $1/p$ and the torque by p. The torque multiplication arises from the derivation of the torque equation, in which an integral is taken round the whole circumference of the rotor; since this comprises $2p$ pole pitches, or p complete cycles of the field, the torque equation becomes

$$T = pkB_{1m}B_{2m} \sin \delta_{12} \tag{4.74}$$

where δ_{12} is the electrical angle between the axes of the stator and rotor fields. The work done per second by the rotating field is $\omega_s T$, and this is independent of the number of poles.

4.8 Introduction to practical windings

For most of this chapter ideal sinusoidally distributed windings have been postulated, in order to simplify the mathematical treatment. One consequence of a sinusoidal distribution is that the rotating field has a constant amplitude B_m and a constant angular velocity ω_s. These are desirable properties, which ensure that the torque developed by the machine will be a steady quantity. Practical windings are designed to approximate to this ideal, and in this section we show what can be achieved with a fairly simple arrangement of conductors in the form of coils.

Consider a two-phase winding. Figure 4.15 shows a distribution of conductors which gives a stepped approximation to the ideal sinusoid for the α phase, and figure 4.16 shows the corresponding field distribution. Conductors carrying i_α inwards are designated α, and those carrying i_α outwards are designated $\overline{\alpha}$. Similar diagrams for the β phase are shown in figures 4.17 and 4.18. These two conductor arrangements can be combined, as shown in figure 4.19, to form a layer of conductors of uniform thickness. In practice the conductors in a machine are placed in slots, as illustrated in figure 4.20. A conductor carrying current inwards in one slot may be joined with another conductor carrying the same current outwards in another slot to form a coil, as shown in figure 4.21. Since one coil side is at the top of a slot and the other side at the bottom of a slot, the process may be continued in this way all round the periphery; all conductors are joined in pairs to form coils, and all coils have the same shape. Figure 4.22 shows coils of this kind fitted into the stator core of a machine. After winding, the coils are interconnected so that all the conductors of one phase are in series (or possibly in series/parallel groups when the machine has more than two poles). It may be observed that this method of constructing an AC stator winding is very similar to the construction of a DC armature winding mentioned in section 2.1.

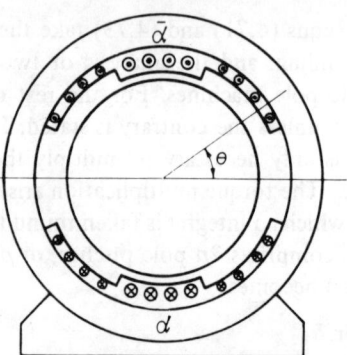

Figure 4.15 Conductor distribution for the α phase

INTRODUCTION TO AC MACHINES 125

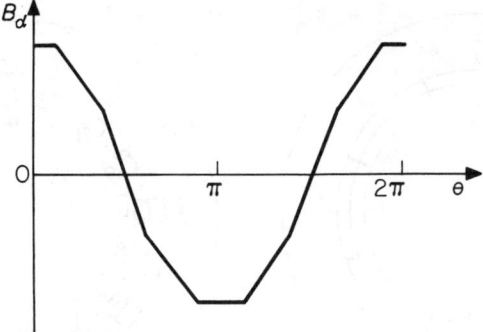

Figure 4.16 Flux density distribution for the α phase

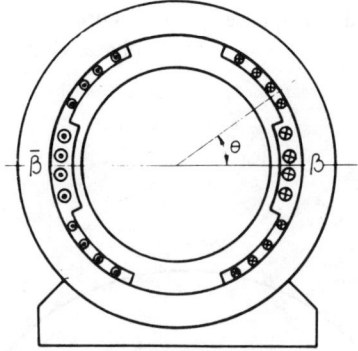

Figure 4.17 Conductor distribution for the β phase

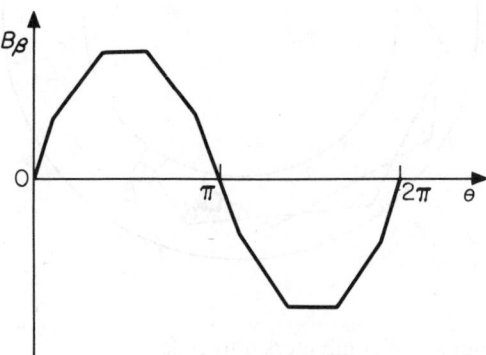

Figure 4.18 Flux density distribution for the β phase

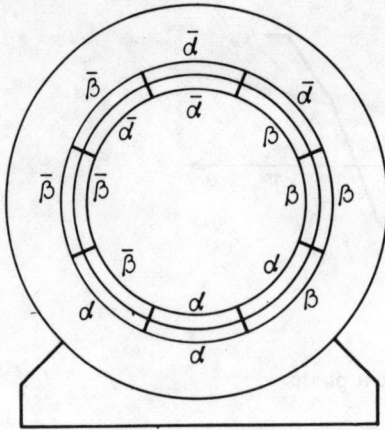

Figure 4.19 Combined two-phase conductor distribution

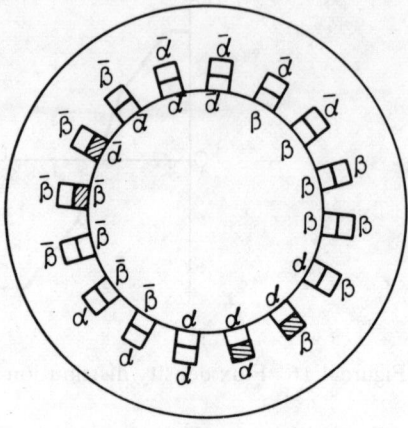

Figure 4.20 Arrangement of a two-phase winding in slots

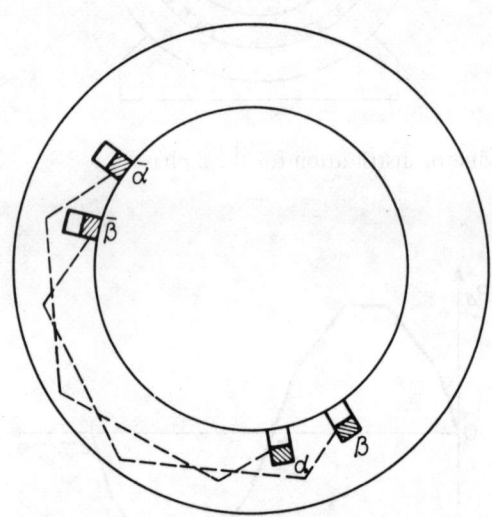

Figure 4.21 Grouping of conductors into coils

Figure 4.22 Partly-wound stator for a 2.6 MW 2-pole induction motor (GEC Large Machines Ltd)

The principle of this two-phase winding can be adapted to three phases, where it gives an even better approximation to the ideal. To see why this is so, consider first a winding consisting of a single concentrated coil as in figure 4.11. If the coil is of negligible width, then eqn (4.3) shows that the resulting field distribution will be a square wave because Σi remains constant for nearly $180°$ and then changes sign as θ passes the position of the next coil side. We may therefore express B in terms of the Fourier series for a square wave of amplitide B_m

$$B = \frac{4B_m}{\pi}(\cos\theta - \frac{1}{3}\cos 3\theta + \frac{1}{5}\cos 5\theta - \ldots) \qquad (4.75)$$

This is the worst possible winding, which introduces a series of unwanted space harmonics in addition to the fundamental $\cos\theta$ term. One advantage of a three-phase winding is that harmonics which are multiples of 3 — the triplen harmonics — usually have little effect. In a three-wire star-connected machine, the triplen harmonics cancel completely, regardless of whether the currents are balanced or not. If the winding is spread over several slots, some of the other harmonics are also attenuated, giving a much improved waveform.

In the two-phase winding discussed above, a non-uniform conductor distribution is achieved by using coils which span less than one pole pitch. Figure 4.20 shows that the span is 3/4 of a pole pitch. For a similar 3-phase winding the span is 5/6 of a pole pitch, and this has the effect of severely attenuating the 5th and 7th space harmonics. For a winding in 24 slots — the three-phase equivalent of figure 4.20 — the Fourier series is as follows, with the triplen harmonics omitted

$$B = \frac{4B_m}{\pi}(0.933\cos\theta + 0.013\cos 5\theta + 0.010\cos 7\theta +$$
$$+ 0.085\cos 11\theta + 0.072\cos 13\theta + \ldots) \qquad (4.76)$$

The space harmonics are now insignificant apart from the 11th and 13th, and these could be reduced by using more slots. This is only one of many possible winding arrangements, but the intricacies of practical windings do not concern us here; they are treated in standard texts such as Say [2]. It is sufficient to know that windings can be made to produce an acceptable approximation to the ideal sinusoidal field, and for the purposes of analysis we assume that the field is exactly sinusoidal.

Problems

4.1. In the definition of the MMF of a distributed winding (section 4.1), the summation of currents is taken from a starting point θ_0 for which $H = 0$. If θ_0 cannot be found by inspection, some arbitrary angle, say $\theta = 0$, must be used as the starting point. Then

$$F(\theta) = F'(\theta) + F(0)$$

where $F'(\theta)$ is the sum of the currents in the conductors between 0 and θ, and $F(0)$ is the MMF at the point $\theta = 0$.

Use the fact that the net flux out of a closed surface is always zero to prove that when the airgap length g is a constant the MMF $F(\theta)$ must satisfy the condition

$$\int_0^{2\pi} F(\theta)\, d\theta = 0$$

Hence show how the value of $F(0)$ may be deduced from a graph of $F'(\theta)$ against θ.

4.2. Extend the analysis of a three-phase winding given in section 4.4 to m phases. Show that the winding will produce a magnetic field rotating with an angular velocity equal to the angular frequency of the supply, and that the amplitude will be constant and equal to $m/2$ times the maximum field produced by any one phase acting alone. Note that the analysis involves the summation of terms of the form $\cos(\omega t + \theta + 4\pi r/m)$ where r is an integer. This is most easily accomplished if the terms are represented by phasors.

4.3. Verify the expression given in eqn (4.55) for the EMF induced in the α-phase winding by considering an elementary coil formed by the conductors in two elementary arcs of angle $d\theta$, one located at $+\theta$ and the other at $-\theta$. If the conductor density is $Z \sin \theta$, calculate the flux linkage of this elementary coil, and by integrating from $\theta = 0$ to $\theta = \pi$ evaluate the total flux linkage of the winding.

4.4. The relationship between the current density and the airgap flux density is given by eqn (4.8) where g is the radial length of the airgap.

Consider a machine with a circular stator, and a stator winding which produces a rotating current wave given by

$$K = K_m \cos(\theta - \omega t - \alpha)$$

The machine has an iron rotor with no windings; it is not circular, but is shaped so that the airgap length is given by

$$1/g = A_0 + A_1 \cos 2\phi$$

where ϕ is the angular displacement from a reference axis on the rotor. If the rotor revolves with an angular velocity ω_r, the position of a point on the rotor with respect to the stator is $\theta = \phi + \omega_r t$, and $1/g$ becomes

$$1/g = A_0 + A_1 \cos 2(\theta - \omega_r t)$$

Find the total torque exerted on the stator winding; this will be equal and opposite to the torque exerted on the rotor. Show that the torque will be constant if $\omega_r = \omega$, and compare this machine with the one considered in problem 1.6.

References

1. P. L. Taylor, *Servomechanisms*, 2nd ed. (London: Longman, 1964).
2. M. G. Say, *Alternating Current Machines*, 5th ed. (London: Pitman, 1983).
3. C. R. Chapman, *Electromechanical Energy Conversion* (New York: Blaisdell Publishing Co., 1965).
4. A. E. Fitzgerald, C. Kingsley, Jr. and S. D. Umans, *Electric Machinery*, 4th ed. (New York: McGraw-Hill, 1983).

5 Synchronous Machines

5.1 Introduction

In its usual form, the synchronous machine consists of a stator with a polyphase winding which produces a rotating magnetic field, and a magnetised rotor having the same number of poles as the stator field. The rotor may incorporate permanent magnets, or it may be magnetised by direct current flowing in a field winding on the rotor. As its name implies, the distinctive feature of the synchronous machine is that the rotor revolves in synchronism with the rotating magnetic field of the stator, and its speed is therefore related to the frequency of the AC supply to the stator.

Synchronous machines can operate as generators or motors; nearly all the large generators in power supply systems are of this kind, and large synchronous motors are widely used as high-efficiency constant-speed industrial drives. Machines of this kind invariably have wound rotors, which permit control of the machine characteristics by varying the rotor excitation. In small machines the DC supply to the rotor is usually taken through brushes and sliprings, but in large machines a brushless excitation system is normally employed as follows. The DC supply to the rotor excitation winding is obtained from a shaft-mounted rectifier supplied by a small AC generator known as an *exciter*; the AC winding of the exciter is mounted on the same shaft as the main machine rotor, and the field winding of the exciter is stationary. Figure 5.1 shows the rotor of a large brushless synchronous generator. With large turbine-driven generators the power loss in the excitation winding is a limiting factor in the design; this loss could be eliminated by using a superconducting winding [1], which offers the prospect of a radically different machine design for ratings above 1000 MW.

Whereas large synchronous motors are normally connected directly to the AC mains supply, small machines can be operated from frequency converters or inverters for speed control by variation of the supply frequency. This provides an alternative to the induction-motor drive systems described in section 6.5. Permanent-magnet synchronous motors are particularly suitable for inverter operation, and new developments [2] claim more power output at a higher efficiency than for an induction motor of the same frame size.

Conventional synchronous machines operate from a three-phase AC supply, and in most cases the supply is the mains with a sinusoidal waveform and a constant frequency. There is a class of motors known as *stepper motors*; they are

132   ELECTRICAL MACHINES

Figure 5.1   Rotor for a 6.2 MVA synchronous generator with brushless excitation (GEC Large Machines Ltd)

related to the conventional synchronous machines, but they operate in quite a different way. Stepper motors are essentially digital machines, controlled by switching the windings in sequence to a DC supply. These machines are growing in importance, and an introduction is given in section 5.4.

This chapter is mainly concerned with the fundamental principles of conventional synchronous machines, and with the use of the excitation winding to control the machine characteristics. An important step in the theory is the development of an equivalent circuit. This concept has already been used with the transformer, to represent the behaviour of magnetically coupled coils by a network of ideal circuit elements (resistors, inductors and an ideal transformer). In a similar way the interaction of currents and magnetic fields in the synchronous machine can be represented by a simple circuit made up of ideal elements, and the characteristics of the machine are readily deduced from the circuit. Thus the equivalent circuit forms a link between the internal electromagnetic processes and the external performance characteristics.

5.2 Phasor diagram and equivalent circuit

We assume that the synchronous machine has a cylindrical rotor, so that it is reasonable to postulate sinusoidally distributed magnetic fields. Figure 5.2 is a space phasor diagram representing the field components at a particular instant of time; the phasors rotate with the synchronous angular velocity ω_s. In this diagram, K_1 represents the current density distribution produced by currents in the

SYNCHRONOUS MACHINES

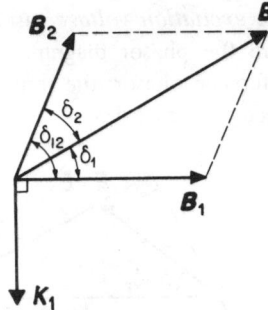

Figure 5.2 Space phasor diagram for an ideal synchronous machine

stator winding, and B_1 is the corresponding magnetic field; B_2 is the magnetic field produced by currents in the rotor winding; and B is the total or resultant magnetic field. The component fields, acting individually, would induce voltages E_1 and E_2 in one phase of the stator armature winding. The total induced voltage E_t is thus equal to $E_1 + E_2$, as shown in the time phasor diagram of figure 5.3.

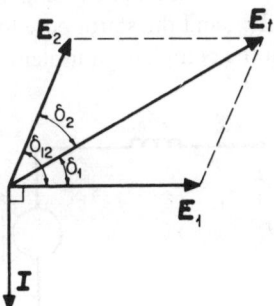

Figure 5.3 Time phasor diagram for an ideal synchronous machine

If the resistance and leakage reactance may be ignored, the total induced voltage E_t will be equal to the terminal voltage V. From eqn (4.57), the component voltage E_1 (due to the stator magnetic field) is related to the stator current by

$$E_1 = j\omega L_m I = jX_m I \tag{5.1}$$

where L_m is the effective inductance of one stator phase, and X_m is the corresponding reactance. The component E_2 is the contribution of the rotor magnetic field to the total induced voltage; since it depends on the rotor excitation current, it is termed the *excitation voltage*, usually denoted by E. With these changes of nomenclature the phasor diagram of figure 5.3 may be redrawn, taking $V = E_t$ as the reference phasor; the result is shown in figure 5.4, where the angle δ is the same as δ_2 in figure 5.3.

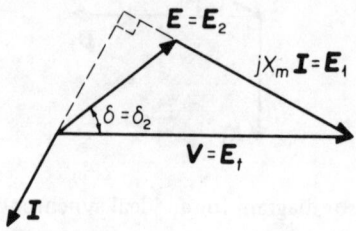

Figure 5.4 Conventional phasor diagram for an ideal synchronous machine

Complete equivalent circuit

Figure 5.4 is the phasor diagram for one stator phase of a synchronous machine; by inspection it is also the phasor diagram for the circuit shown in figure 5.5, which is therefore an *equivalent circuit* of one phase of the machine. The resistance and leakage reactance of the armature winding may be included as elements in series with the reactance X_m, and the stator core loss may be represented by a shunt resistance R_c, to form the complete equivalent circuit shown in figure 5.6.

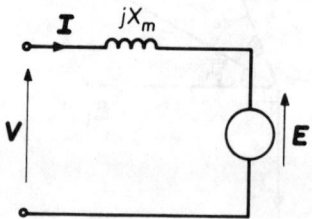

Figure 5.5 Equivalent circuit of an ideal synchronous machine

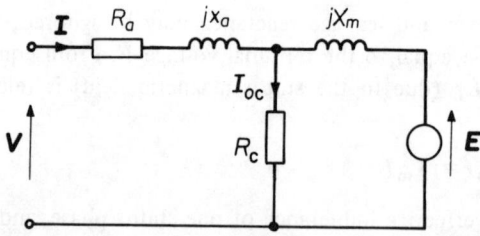

Figure 5.6 Complete equivalent circuit of the synchronous machine

It is useful to transform the complete equivalent circuit so that the reactance X_m appears as a shunt element; this permits a direct comparison with the equivalent circuit of the induction machine which is derived in chapter 6. The transformation is accomplished by replacing the series combination of E and jX_m with an equivalent parallel circuit, as shown in figure 5.7. The complete equivalent circuit for the synchronous machine then takes the form shown in figure 5.8, and the resemblance to the equivalent circuit of the transformer will be noted. In this circuit (figure 5.8) the reactance X_m is a mutual or magnetising reactance, with the magnetising current I_{0m} supplied partly by the stator current and partly by the rotor excitation. The flux associated with this reactance represents the total magnetic field in the machine airgap, which is the resultant of fields produced by the stator and rotor currents.

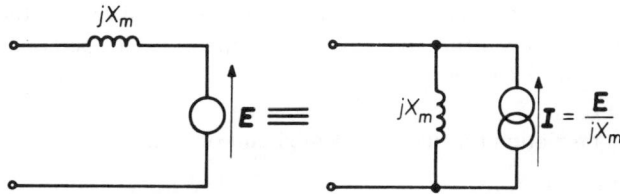

Figure 5.7 Transformation of the excitation branch

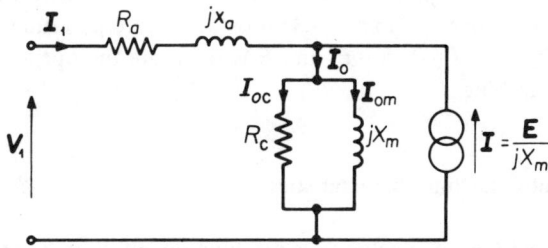

Figure 5.8 Transformed equivalent circuit of the synchronous machine

Approximate equivalent circuit

The armature resistance R_a is often small in comparison with the reactance X_m, and it may be a reasonable approximation to ignore it. If the core loss resistance R_c is also ignored, the two reactances x_a and X_m in figure 5.6 may be combined to form a single reactance X_s, known as the *synchronous reactance*. The equivalent circuit then takes the simple form shown in figure 5.9, which is quite a good

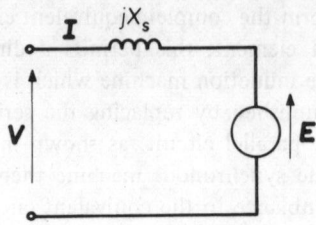

Figure 5.9 Approximate equivalent circuit of the synchronous machine

representation of large machines. With small machines the resistance R_a can be more than 10 per cent of the synchronous reactance X_s, and it should be included in series with X_s if greater accuracy is required. The omission of the core loss resistance R_c has very little effect for most conditions of operation, since the current I_{oc} (figure 5.6) is very small in comparison with the normal value of the total current I.

The phasor equation for the equivalent circuit of figure 5.9 is

$$V = E + jX_s I \qquad (5.2)$$

and when the resistance R_a is included this becomes

$$V = E + (R_a + jX_s)I \qquad (5.3)$$

The representation of eqn (5.2) by a phasor diagram will be considered in the next section. Equations (5.2) and (5.3) provide an interpretation of the excitation voltage E; when $I = 0$, $V = E$, and E is therefore the open-circuit terminal voltage of the machine.

5.3 Synchronous machine characteristics

The essential features of synchronous machine operation may be derived from the approximate equivalent circuit and the corresponding phasor diagram; before doing so, however, it is useful to make some general deductions from the rotating field concepts.

Synchronous speed

It was shown in section 4.7 that the angular velocity of the rotating magnetic field is given by $\omega_s = \omega/p$, where ω is the angular frequency of the supply and p is the number of pole pairs. If the frequency of the mains supply is f hertz, it follows that the synchronous speed of the machine will be f/p rev/s, or $60f/p$ rev/min. With a mains frequency of 50 Hz the speed of a two-pole machine will be 3000 rev/min, the speed of a four-pole machine will be 1500 rev/min, and so on. Once the number of poles has been selected by the designer, the speed of a synchronous machine can be altered only by varying the supply frequency. The synchronous machine is therefore a constant-speed device when operated directly from the AC mains supply.

Synchronous torque

When the rotor is running at the synchronous speed, its poles will be displaced by a constant angle from the effective poles of the stator. The magnetic lines of force at a particular instant of time are as shown in figure 4.8, and this magnetic field pattern rotates without changing its form. A constant electromagnetic torque is therefore exerted on the rotor, and this will balance the mechanical torque applied to the shaft. The torque equation deduced for stationary fields may therefore be applied to the synchronous machine, giving

$$T = kB_{2m}B_m \sin \delta_2 \qquad [4.30]$$

Now $B_m \propto V$, $B_{2m} \propto E$ and $\delta_2 = \delta$; the torque equation becomes

$$T \propto VE \sin \delta \qquad (5.4)$$

and if the terminal voltage V and excitation voltage E are held constant then

$$T \propto \sin \delta$$

Load angle

With the convention that positive values of T and δ correspond to motoring operation, the variation of torque with the angle δ is as shown in figure 5.10. This is an important characteristic of the synchronous machine, and the angle δ is termed the *load angle*; its value varies according to the load applied to the

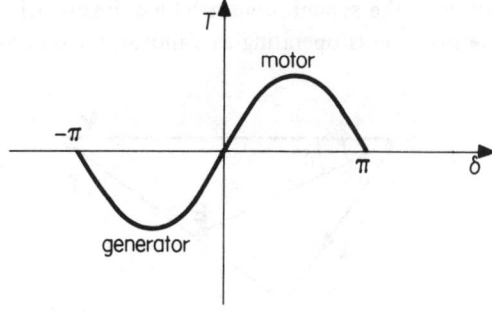

Figure 5.10 Torque/load-angle characteristic for the synchronous machine

shaft, within the limits of $\pm\pi/2$ radians. Since the speed of the machine is fixed, the torque is entirely determined by the mechanical system connected to the machine shaft. If this is a mechanical load, the torque T will be positive; the synchronous machine will act as a motor, and the positive value of δ implies that the stator poles are ahead of the rotor poles. If a source of mechanical power is coupled to the shaft, T will be negative and the synchronous machine will act as a generator. The angle δ will also be negative, showing that the stator poles are now lagging behind the rotor poles. The magnitude of the torque has a maximum value, known as the *pull-out torque*, when $\delta = \pm\pi/2$ radians. If a torque in excess of this value is applied to the machine shaft, the electromagnetic torque cannot balance the shaft torque, and synchronism will be lost.

Damper windings

In the normal working range, the torque/load-angle characteristic is approximately linear; the characteristic resembles that of a spring, where torque increases with displacement. Synchronous machines thus exhibit an electromagnetic 'springiness', and the rotor will tend to oscillate about a mean position if there is any disturbance such as a sudden change in the load. Such oscillations are undesirable, and must be damped; this is normally done electromagnetically, by providing short-circuited windings on the rotor. These windings are known as *damper windings*; they may take the form of conducting paths in solid iron poles or they may be constructed from bars set in slots in laminated poles, like the rotor cage of an induction motor (see chapter 6). During normal synchronous running the flux through a damper winding is constant, and it has no effect; but a change in the load angle will cause the flux to change, and the induced currents will produce a torque which opposes the change.

Synchronous motors and generators

As with the DC machine, there is no essential difference between motoring and generating operation of the synchronous machine. Figure 5.11 shows the phasor diagram when the machine is operating as a motor; the terminal voltage V leads

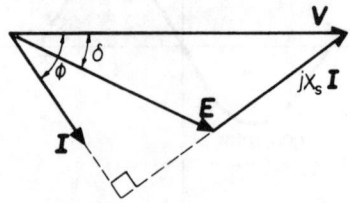

Figure 5.11 Phasor diagram for motor operation

the excitation voltage E by the load angle δ, and the armature current I therefore has a component in phase with V. When the shaft torque is reversed, so that the machine is driven as a generator, the phasor diagram takes the form shown in figure 5.12. The voltage V lags E by the load angle δ, and the current I has a component in antiphase with V.

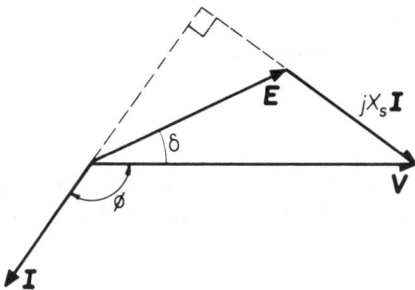

Figure 5.12 Phasor diagram for generator operation

Suppose that the shaft torque gradually changes from a positive (motoring) value to a negative (generating) value, with the magnitudes E and V held constant; the phasor diagram will gradually change from the form shown in figure 5.11 to the one shown in figure 5.12. The locus of E will be an arc of a circle with its centre at O, as shown in figure 5.13, and the locus of I will be another arc with its centre at O'. When the torque is zero, the load angle δ will be zero; I will be in quadrature with V, and the electrical power will also be zero.

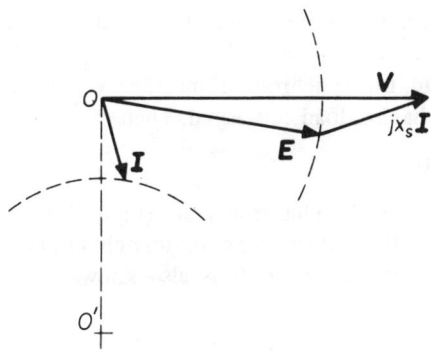

Figure 5.13 Locus diagram for constant excitation

Synchronous motor characteristics

It has already been mentioned that the ability to vary the rotor (or field) excitation is an important feature of the synchronous machine, and we now consider the effect of such a variation when the machine operates as a motor with a constant load. Similar results hold for the synchronous generator with constant mechanical input power.

When the torque load on the motor is constant the power output will be constant, and if losses are neglected there will be a constant input power per phase given by

$$P = VI \cos \phi \tag{5.5}$$

If the voltage V is constant this equation implies that $I \cos \phi$ is constant, and the locus of the current phasor I is the line AB in figure 5.14. From this diagram we have

$$MN = E \sin \delta = X_s I \sin(\pi/2 - \phi)$$

$$= X_s I \cos \phi \tag{5.6}$$

Thus $E \sin \delta$ is a constant, and the locus of E is the line CD. If $\phi = 0$, the machine operates at unity power factor and I has a minimum value; let $E = E_0$ for this condition. When $E < E_0$, ϕ is negative and the machine takes a lagging current, as shown in figure 5.14; the machine is said to be *under-excited*, and synchronism will be lost when $\delta = \pi/2$. When $E > E_0$, the machine is *over-excited*; ϕ is positive and the machine takes a leading current (figure 5.15). If the current I is plotted against the excitation voltage E for different values of the power P, the result is a set of curves known as *V*-curves (figure 5.16). A useful characteristic of the synchronous motor is the leading phase angle of the current when the machine is over-excited. It can be used to compensate for the lagging current taken by other loads such as induction motors, so that the total load power factor is unity. Thus in figure 5.17, if the lagging load current is I_l and the synchronous motor current I_s is suitably adjusted by controlling the excitation, the total current I will be in phase with the voltage V. This is known as *power-factor correction*, and the synchronous machine can be used solely for this purpose, with no mechanical load connected. Then

$$VI \cos \phi = P \to 0 \tag{5.7}$$

therefore $I \cos \phi \to 0$, and the phasor diagram (figure 5.15) shows that $\phi = \pi/2$. Since the current leads the voltage by approximately $\pi/2$ radians, the machine is known as a *synchronous capacitor*; it is also known as a *synchronous compensator*.

SYNCHRONOUS MACHINES 141

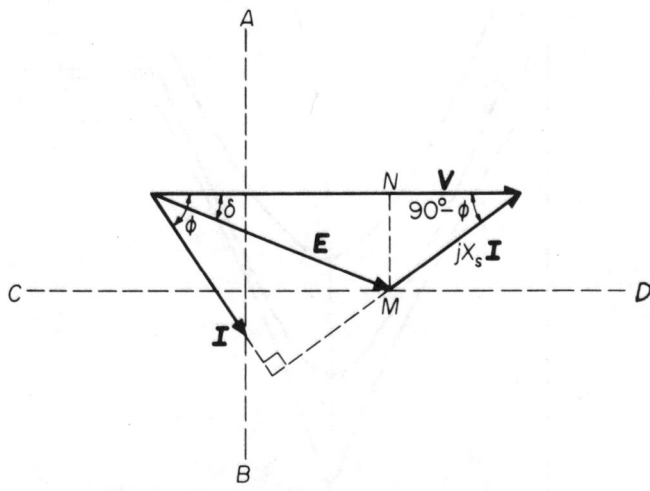

Figure 5.14 Locus diagram for constant power

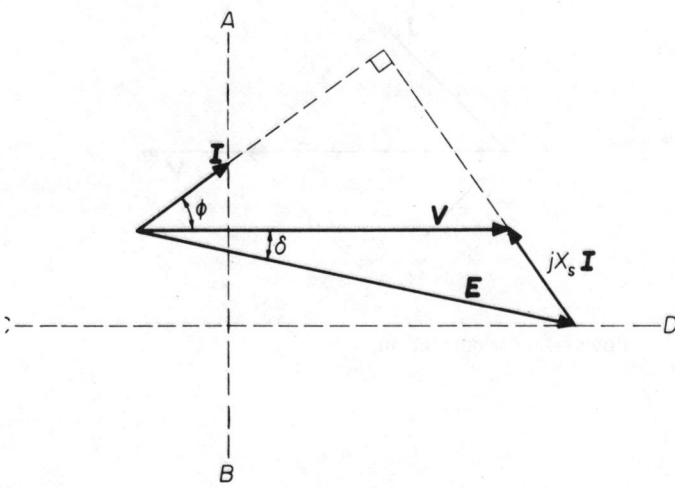

Figure 5.15 Leading power-factor condition

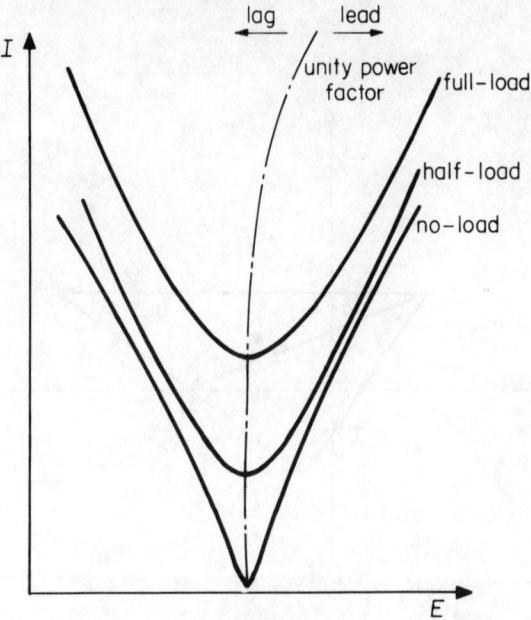

Figure 5.16 Synchronous motor V-curves

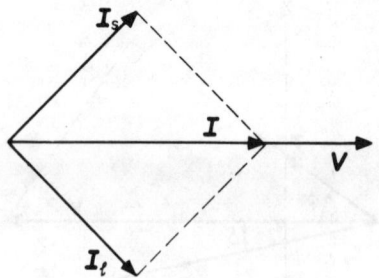

Figure 5.17 Power-factor correction

Torque equation

The torque equation for the synchronous motor may be deduced from the phasor diagram. Let $\omega_s = \omega/p$ be the synchronous angular velocity; the total mechanical output power is $\omega_s T$, and if there are m phases the power per phase

SYNCHRONOUS MACHINES

is $\omega_s T/m$. If losses are neglected this must equal the electrical input power per phase, so that

$$\frac{\omega_s T}{m} = P = VI \cos \phi \tag{5.8}$$

Substituting for $\cos \phi$ from eqn (5.6) gives

$$T = \frac{m}{\omega_s X_s} VE \sin \delta \tag{5.9}$$

This agrees with eqn (5.4), which was derived from a consideration of the magnetic fields.

Starting of synchronous motors

When the rotor is running at less than the synchronous speed, the load angle δ will increase continuously. Figure 5.10 shows that the torque T will be alternately positive and negative, with a mean value of zero. A synchronous motor is therefore not inherently self-starting; some other principle must be used to accelerate the rotor to a speed close to synchronism. Once the speed is high enough, the motor will synchronise; the first positive half-cycle of torque will pull the rotor into step with the rotating field.

Induction motors develop a positive torque at speeds down to zero, and the induction motor principle is generally used for starting synchronous machines. Occasionally a small induction motor (known as a 'pony motor') is coupled to the synchronous machine shaft; this motor is connected to the AC supply for the starting process, and disconnected once the main machine rotor is synchronised. More commonly, the synchronous machine rotor itself is used; the field winding is disconnected from the DC excitation source, and either short-circuited or connected to a starting resistor. When an AC supply is connected to the stator, currents will be induced in the rotor field and damper windings, resulting in a torque which accelerates the rotor.

5.4 Salient-pole synchronous machines

The theory developed in this chapter is only valid for machines with a uniform airgap, that is, those in which the rotor and the stator bore are both cylindrical. When the rotor has salient poles (as shown in figure 5.18) the reactance is no longer constant. Consider first the magnetic field conditions in a non-salient machine, represented by figure 5.2. When there is no load on the machine, $\delta_{12} = 0$, and the magnetic field B_1 due to stator currents acts in the direction of the N–S pole axis of the rotor. If the load is increased until $\delta_{12} = \pi/2$, then the field B_1 is at right angles to the rotor pole axis. In a salient-pole machine, the

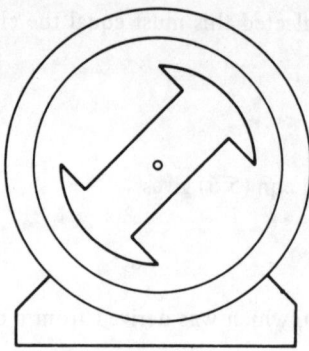

Figure 5.18 Salient-pole synchronous machine

condition $\delta_{12} = 0$ gives a path of easy magnetisation (low reluctance) along the rotor pole axis; this is termed the *direct axis,* and the corresponding reactance at the machine terminals has a high value X_d. The condition $\delta_{12} = \pi/2$ gives a path of difficult magnetisation (high reluctance) along an axis at right angles to the rotor pole axis; this is termed the *quadrature axis,* and the corresponding reactance at the machine terminals has a low value X_q. It is no longer possible to represent the machine by a simple equivalent circuit, and the torque equation contains an additional term. An outline of the theory is given in section 7.5, and the torque equation developed there for a two-pole three-phase machine may be generalised for $2p$ poles and m phases

$$T = \frac{mp}{\omega}\left[\underbrace{\frac{VE}{X_d} \sin \delta}_{(a)} + \underbrace{\frac{V^2}{2}\left\{\frac{1}{X_q} - \frac{1}{X_d}\right\} \sin 2\delta}_{(b)}\right] \qquad (5.10)$$

Term (a) in this equation represents the normal synchronous torque, which may be identified with eqn (5.9). Term (b) represents a component of torque due to the saliency of the rotor; this component is termed the *reluctance torque,* and it vanishes when the rotor is cylindrical, for X_d is then equal to X_q.

A qualitative explanation of the origin of the reluctance torque is that the salient poles of the rotor will tend to line up with the axes of the stator magnetic field. The torque will be zero when the rotor pole axis is in line with the stator field axis, and zero again when the rotor pole axis is at right angles to the stator field axis. In contrast, the synchronous torque will have a maximum or a minimum value when the two axes are at right angles. The reason for the difference is that the reluctance torque is dependent on the induced magnetisation of the rotor, which varies with the rotor position; whereas the synchronous torque depends on the rotor magnetisation produced by the field winding, and this is independent of the rotor position. Thus in eqn (5.10), the reluctance torque varies as $\sin 2\delta$, while the synchronous torque varies as $\sin \delta$. The torque/load-angle characteristic for a salient-pole machine therefore takes the form shown in figure 5.19.

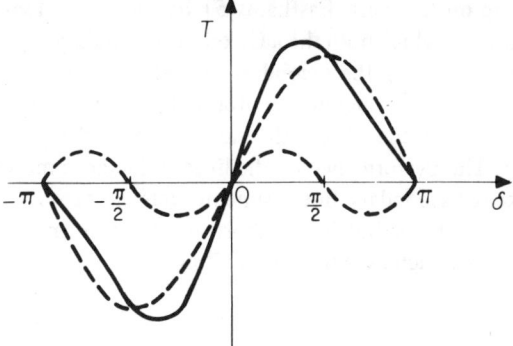

Figure 5.19 Torque/load-angle characteristic for a salient-pole synchronous machine

Reluctance motors

An interesting possibility is to omit the rotor field winding from a salient-pole synchronous motor, so that only the reluctance torque term remains in eqn (5.10). Machines of this kind are known as *reluctance motors* [3], and they normally incorporate on the rotor an induction starting winding of the cage type (see chapter 6). For good performance the rotor must be designed to make the reactance ratio X_d/X_q as large as possible; the simple salient-pole construction is not the best form of the rotor, and commercial motors use other devices such as flux barriers or segmentation to improve the performance [3]. Reluctance motors combine the advantages of the cage induction motor with the speed characteristic of the synchronous machine. They are used in place of induction motors when a constant speed, related to the supply frequency, is required. A typical application is to maintain an absolute speed relationship between a number of widely separated shafts, where mechanical coupling would be difficult.

Linear synchronous motors

Linear induction motors are described in section 6.7. Synchronous motors can also be made to produce linear motion by changing the geometry from a cylindrical structure to a flat structure, so that the AC winding generates a travelling magnetic field. The part which carries the AC winding is termed the *primary*, and the linear counterpart of the field system is termed the *secondary*; as with the linear induction motor, a linear synchronous motor is made with either a short primary or a short secondary. One possible application of such machines is in advanced passenger transport systems [4]. If the secondary forms part of the track for the vehicle, it is undesirable to have either permanent magnets or an excitation winding on this part. Eastham [5] has described two types of linear synchronous motor in which both the DC excitation winding and the AC winding are on the primary, leaving the secondary entirely passive. An alternative is to use a linear form of the reluctance motor [6], which requires no secondary excitation; the primary then takes the same form as the primary of a linear induction motor. The performance of the linear reluctance motor is not as good as that of an excited-secondary linear synchronous motor, so it is unlikely to be used for transport systems; but in the short-secondary form it has potential for industrial applications such as materials handling.

5.5 Stepper motors

The synchronous motors considered so far in this chapter have two features in common: the AC armature winding is effectively sinusoidally distributed, and the AC supply is taken from the three-phase mains. In consequence a synchronous motor develops a smooth torque and it runs at a constant speed. There is an important class of motors known as *stepper motors* or *stepping motors*; they are related to synchronous motors in the sense that the motion of the rotor is determined by the frequency of currents in the armature windings, but they differ in two important respects. Firstly, the armature windings are not even approximately sinusoidally distributed; secondly, the supply is not sinusoidal AC but switched DC. As its name implies, the function of a stepper motor is to move the rotor through a precise angular step when the current in one or more of the stator windings is switched. The theory of the stepper motor is complex, and is beyond the scope of this book; only a qualitative treatment can be given in this section, and the reader should consult Acarnley [7] for an introduction to the theory.

In a stepper motor the windings are supplied from semiconductor drive circuits – usually employing transistors – which in turn are controlled by a 'translator' that switches the currents sequentially in response to successive drive pulses. Thus each pulse input to the translator causes a switched change in the motor winding currents, which in turn should move the rotor through one step. If pulses are applied at a low repetition rate, the winding currents will remain constant between the pulses and the rotor will come to rest after each step. At high stepping rates the rotor will move continuously, though its velocity will not be constant, and the motor behaves like a conventional synchronous motor.

The correspondence between rotor steps and input pulses makes the stepper motor an ideal device for digital control. Applications include quartz analogue clocks and watches, printers and graph plotters, head positioning in computer disc drives, and numerically controlled machine tools. They are made in sizes ranging from milliwatts to tens of kilowatts [7], and they are expected to replace conventional AC and DC machines in many control applications.

Permanent-magnet stepper motors

The permanent-magnet stepper motor is similar to an ordinary synchronous motor with a permanent-magnet rotor. It differs in the structure of the stator winding, which takes the form of coils wound on salient poles. Energising the windings on one pair of poles will pull the rotor into alignment with those poles, and energising the next pair will move the rotor through one step. The torque per unit volume is relatively poor with this type of motor, and the step angle is large: $90°$ in the case of a two-pole rotor. In most applications larger torques and smaller step angles are required, so the permanent-magnet stepper motor is not widely used.

Variable-reluctance stepper motors

The variable-reluctance stepper motor is related to the reluctance motor described in section 5.3, in that the rotor carries no windings or permanent magnets, and the torque production depends on the change in the reluctance of a magnetic circuit with rotor position. Both the stator and the rotor have salient poles or teeth, and the number of rotor poles may be different from the number of stator poles.

Figure 5.20 shows a cross-section through a simple variable-reluctance motor. There are three stator phases, arranged on six poles, and four rotor poles. The diagram shows three successive positions of the rotor when phases a, b and c are energised in sequence; the step angle is $30°$. Smaller step angles can be achieved by using more stator and rotor poles, or by employing a multi-stack construction.

In the multi-stack motor [7] there is a separate stator and rotor combination for each phase. The rotor has small salient poles or teeth, and the stator usually has four poles with similar teeth formed in the pole faces. When one phase is energised, all the rotor teeth in that stack are pulled into alignment with the stator teeth. The rotors are mounted on a common shaft, and the stators in successive stacks are displaced by an angle equal to the step angle. Switching the current from one phase to the next will pull the next set of rotor teeth into alignment with the corresponding stator teeth, causing the shaft to rotate through one step. Multi-stack motors usually have at least three phases.

Variable-reluctance stepper motors have the useful property that the torque depends on the magnitude of the current in a phase and not its direction; the magnetic alignment force is not affected by a reversal of the field direction. Consequently the current in a phase only needs to be switched on and off; it does not need to be reversed. This greatly simplifies the electronic control of variable-reluctance motors.

148 ELECTRICAL MACHINES

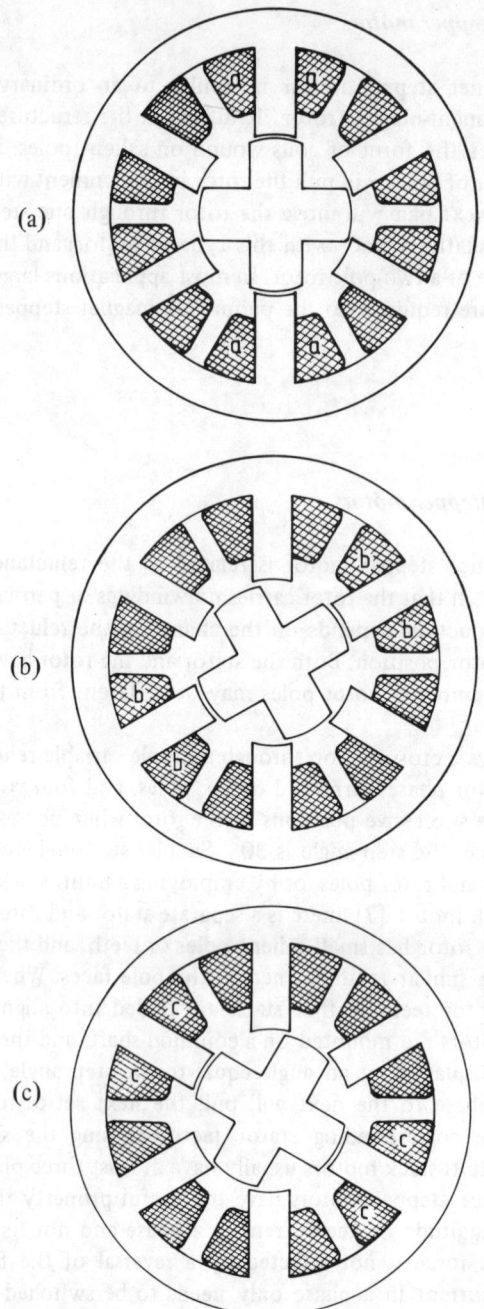

Figure 5.20 Single-stack variable-reluctance stepper motor: (a) phase a energised; (b) phase b energised; (c) phase c energised

Hybrid stepper motors

A permanent magnet can be used to augment the variable-reluctance effect, resulting in a stepper motor with the highest torque per unit volume [7]. The stator is similar to that of the variable-reluctance motor, but the rotor is made in two parts which form the N and S pole-pieces of a permanent magnet. Figure 5.21 illustrates the principle of the two-phase motor. Figure 5.21(a) shows a longitudinal section through the motor, and figure 5.21(b) shows transverse sections through the rotor N and S poles. Unlike the multi-stack variable-reluctance motor, the hybrid motor has a single stator with long poles bridging the two rotor parts, and there is an angular displacement between the teeth on the two rotor parts. Coils on stator poles 1 and 3 form the α phase winding, and coils on poles 2 and 4 form the β phase. Figure 5.21(b) shows the condition when phase α is energised with positive current. The stator flux opposes the permanent-magnet flux in gaps 3-N and 1-S, but it aids the permanent-magnet flux in gaps 1-N and 3-S; the rotor teeth are pulled into alignment with the stator teeth in the gaps where the field is strongest. Figure 5.21(c) shows the corresponding condition when phase β is energised, so that the teeth are aligned in gaps 2-N and 4-S. Negative current in phase α will cause alignment in gaps 3-N and 1-S, and negative current in phase β will cause alignment in gaps 4-N and 2-S. Thus the current sequence $+\alpha, +\beta, -\alpha, -\beta$ will result in a rotor movement of one tooth pitch. Practical two-phase motors usually have eight stator poles and more teeth — typically 50 rotor teeth, with 5 teeth on each stator pole, giving a step angle of $1.8°$.

Electronic drive systems

Unipolar and bipolar drives

Electronic drive systems for stepper motors are broadly classified as unipolar or bipolar. Unipolar drives supply current in one direction only; bipolar drives can reverse the direction of current. Variable-reluctance stepper motors are well suited to unipolar drives, because the current in a phase only needs to be switched on and off. Motors incorporating permanent magnets, on the other hand, require the field in each phase to be reversed from time to time. This can be achieved by reversing the current in a single winding with a bipolar drive, or by switching the current in bifilar windings as follows: each pole is wound with two coils in opposite directions, so that the field can be reversed by passing current of the same polarity through alternate coils. Bipolar drives are considerably more complicated than unipolar drives; a simple bipolar drive uses four power transistors per phase, arranged in a bridge circuit, whereas a simple unipolar drive requires only a single transistor per phase [7]. A motor with bifilar windings can

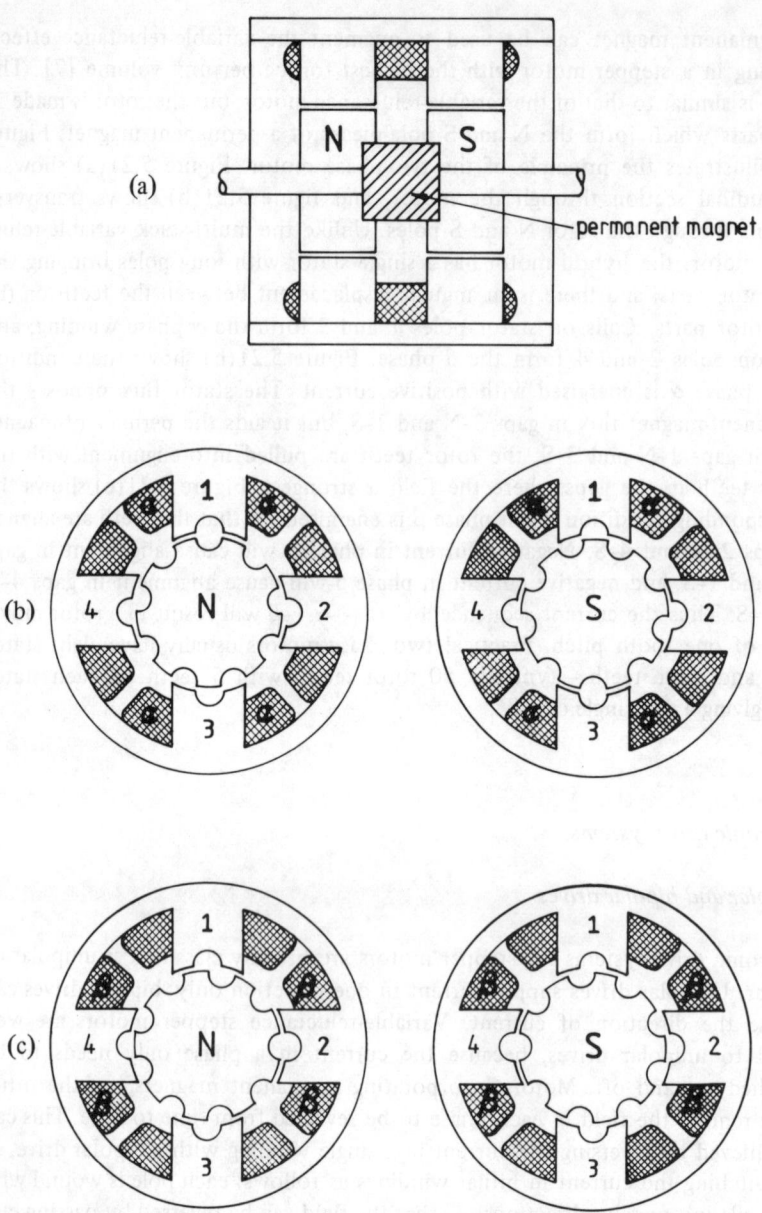

Figure 5.21 Hybrid stepper motor: (a) longitudinal section; (b) transverse section through rotor poles: phase α energised; (c) phase β energised

be supplied from a unipolar drive, but there is a performance penalty: only half the space is available for each active winding, so the effective resistance per phase is doubled.

Current control

In the simplest drive circuits, the power transistor simply acts as a switch between the motor phase winding and a low-voltage supply. In the steady state the current is limited by the resistance of the winding, and the supply voltage is chosen to give the rated current under these conditions. When the switch is turned on, the rate of rise of current is determined by the time constant L_p/R_p of the phase winding, and this limits the speed of operation of the motor.

The speed of response of the simple drive can be improved by voltage forcing; a high supply voltage is used, and the current is limited by connecting a forcing resistance R in series with the winding. The time constant is now $L_p/(R_p + R)$, and if R is several times larger than R_p then the current rise-time will be correspondingly shortened. Although simple, this system is inefficient because of the energy lost in the forcing resistance. A better system is the chopper drive, which gives the benefits of forcing without the energy loss.

In the chopper drive, a high-voltage supply is connected to the motor winding until the current rises above the rated value; the supply is then disconnected until the current falls below the rated value, and the cycle repeated. The current rise-time can be made short by using a high enough voltage, and there is no forcing resistance to dissipate energy. Although it uses more components, the chopper drive is now the preferred system where high performance and good efficiency are required.

Stepper motor characteristics

Static torque characteristics

When constant current is passed through one or more phases of a stepper motor, the rotor will experience a torque when it is displaced from its equilibrium step position. This torque tends to restore the rotor to its equilibrium position: a positive displacement will give a negative torque, and vice versa. To a first approximation, the torque of a variable-reluctance or hybrid stepping motor is given by [7]

$$T = -T_m \sin n\theta \tag{5.11}$$

where θ is the angular displacement of the rotor, n is the number of teeth on the rotor, and T_m is the peak static torque. This has the same form as the synchronous machine torque/load-angle characteristic shown in figure 5.10, with δ replaced by $n\theta$.

If the rotor is subjected to an external torque load, it will move from its equilibrium step position until the restoring torque is equal and opposite to the applied torque. The static torque characteristic repeats with a period of one rotor tooth pitch, when θ increases by $360/n$ degrees. Maximum torque is attained with a displacement of a quarter of a tooth pitch, or $90/n$ degrees, and this represents the static stability limit. If the static torque load is gradually increased, the displacement angle will increase until the stability limit is reached. Any further increase in the displacement will take the operating point into the unstable part of the characteristic, where the torque decreases with increasing angle, and the rotor will slip a tooth — it will move to the next equilibrium position one tooth pitch away. If the equilibrium step position is the desired position of the rotor, then a static load can displace the rotor by $\pm 90/n$ degrees from this position before tooth slipping occurs. The maximum positional error is therefore $\pm 90/n$ degrees; this is equal to the step angle in a two-phase hybrid motor or a four-stack variable-reluctance motor.

The peak static torque T_m is related to the current in the stator phases. In the absence of magnetic saturation, T_m varies as the square of the current in a variable-reluctance motor, and directly as the current in a hybrid motor [7]. In practice there is considerable saturation, and the variation of torque with current must be obtained from manufacturers' data.

Multi-step operation

If a single pulse is applied to a stepper motor drive unit, the phase currents will switch to the next step position and the rotor will move to a new equilibrium position. The torque/displacement characteristic is a spring-like characteristic, where the restoring torque increases with displacement; consequently the rotor will oscillate about its new equilibrium position. When repetitive pulses are applied at a low rate, the rotor will move in steps but it will oscillate about each step position. If the next drive pulse is applied before the oscillation has decayed to a low level, then resonance can occur, and the oscillation amplitude may grow until the motor operation becomes erratic [7]. Resonance can occur at stepping rates such that the time between drive pulses is a multiple of the rotor oscillation period; the critical stepping rates are thus given by

$$f_r = f_n/k, \quad k = 1, 2, 3, \ldots \tag{5.12}$$

where f_r is the resonant stepping rate and f_n is the rotor natural frequency of oscillation. Acarnley [7] describes methods of damping the rotor oscillation.

High-speed operation

When the stepping rate is higher than the rotor natural frequency, the motion of the rotor is continuous and the small speed variations between steps can be ignored. This mode of operation is known as 'slewing', and it resembles the normal operation of a conventional synchronous motor.

Provided that the stepping rate is not too high, a stepper motor will start and run synchronously when a train of drive pulses is suddenly applied; it will also stop suddenly when the pulse train stops. There is a critical stepping rate known as the *start/stop rate*; below this rate a stepper motor will start and stop with the pulse train, and the number of rotor steps will equal the number of drive pulses; above this rate the rotor may fail to accelerate, or it may run on for several steps when the pulse train stops. The start/stop rate depends on the total rotor inertia as well as the properties of the motor and the drive unit.

As with a synchronous motor, a slewing stepper motor will stall when the applied torque exceeds a value known as the *pull-out torque*. The pull-out torque varies with stepping rate; an explanation is as follows. Figure 5.22 shows the measured pull-out torque characteristic for a typical small hybrid stepper motor supplied from a chopper drive unit. At low stepping rates the drive unit forces the currents in the motor phases to be almost rectangular pulses of constant amplitude, and the pull-out torque is nearly constant. Since the phases are inductive, it takes time for the current to rise to its full value; this time will form a larger proportion of the step time at higher stepping rates, so the mean

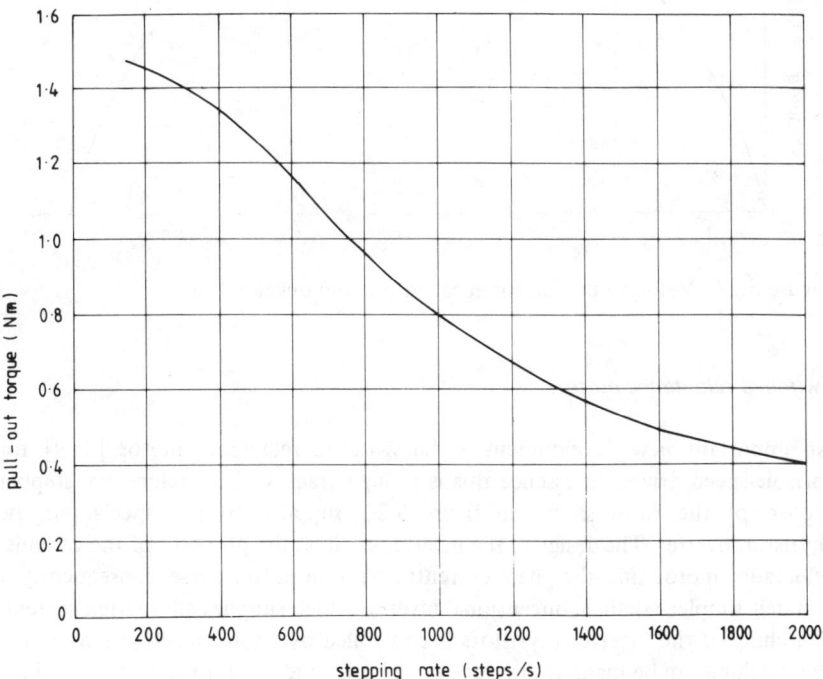

Figure 5.22 Pull-out torque characteristic for a hybrid stepper motor

current will fall and the pull-out torque will decrease. At still higher stepping rates the step time will be too short for the current to reach its full value, and the mean current will be further reduced.

If a stepper motor is to be run at a speed above the start/stop rate, then it is necessary to start the motor at a lower speed and accelerate it by increasing the stepping rate progressively. The converse is required to stop the motor in a controlled manner. A graph of stepping rate against time is termed a *velocity profile*, and figure 5.23 shows a typical profile for starting and stopping a motor. If the load torque is neglected, the maximum torque available for acceleration or deceleration is somewhat less than the pull-out torque; the acceleration thus falls with increasing stepping rate, and this is the reason for the non-linearity in figure 5.23. If the available torque falls linearly with increasing stepping rate, then the required shape for the velocity profile is exponential. A straight line is often a good approximation to the torque characteristic, so an exponential velocity profile is widely used.

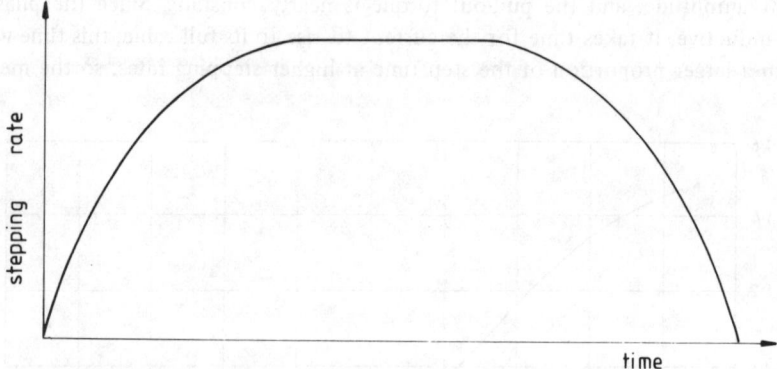

Figure 5.23 Velocity profile for acceleration and deceleration

Switched reluctance motors

An important new development is the switched reluctance motor [8, 9] for variable-speed drives. In essence this is a single-track variable-reluctance stepper motor, of the form shown in figure 5.20, supplied from a special-purpose thyristor inverter. The design of the inverter exploits the property of the variable-reluctance motor that the phase currents do not need to reverse; consequently it is much simpler than a conventional inverter which supplies alternating current. Switching of the inverter thyristors is controlled by a rotor position sensor, and the machine can be made to produce negative torque by control of the switching angle. It is claimed [9] that a switched reluctance motor can produce up to 40 per cent more power than an induction motor of the same frame size, and that the overall efficiency of the drive system (motor and inverter) is better than that of an induction motor alone. Disadvantages include torque pulsation, which can be troublesome at low speeds, and noise.

In a switched reluctance drive, the machine and the inverter are not designed independently because the behaviour of one part is strongly dependent on the design of the other. This drive system marks a new departure in the relationship of machines and power electronics, which deliberately exploits the non-sinusoidal waveforms that switching circuits generate. Conventional AC and DC drives are based on established machines, and the designers of their power electronic controllers try to generate the sinusoidal AC or smooth DC that such machines require.

Problems

5.1. A three-phase star-connected synchronous generator has a reactance of 10 Ω per phase, and it operates with a constant line voltage of 520 V. When the generator is delivering its normal rated power, the line current is 40 A and the power factor is unity. Calculate the output power and the magnitude of the excitation voltage per phase under these conditions.

With the excitation voltage unchanged, the output power of the generator is increased to its maximum value. Calculate the new values of line current, power factor and output power.

5.2. In the steady state a synchronous motor operates with a load angle δ_0 and it delivers a torque T_0 to the load. A momentary fall in the supply voltage causes the load angle to increase by a small amount, and after the voltage disturbance has passed the load angle is given by

$$\delta = \delta_0 + \epsilon$$

If the electromagnetic torque is then given by

$$T = T_0 + \Delta T$$

show that

$$\Delta T = \frac{T_0}{\tan \delta_0} \epsilon$$

It may be assumed that the machine has a cylindrical rotor; that the steady-state equations are applicable; and that the armature resistance may be neglected.

5.3. If the torque load on the motor in problem 5.2 remains unchanged, show that the rotor equation of motion is

$$\frac{J}{p} \frac{d^2 \epsilon}{dt^2} + \frac{T_0}{\tan \delta_0} \epsilon = 0$$

where J is the moment of inertia of the rotor and p is the number of pole pairs. Hence show that there will be small oscillations superimposed on the steady motion of the rotor, and find the frequency of the oscillations.

5.4. With the machine considered in problem 5.3, show that there will be an alternating voltage induced in the rotor field winding, and explain what effect this might be expected to have on the motion of the rotor.

5.5. In chapter 7 it is shown that the general equations for a synchronous machine reduce to the following form when the rotor is cylindrical and the armature resistance is neglected

$$V \sin \delta = X_s I \cos(\delta - \phi)$$

$$V \cos \delta = E - X_s I \sin(\delta - \phi)$$

Show that these equations are satisfied by the phasor diagram of figure 5.11.

References

1 J. L. Smith, 'Overview of the development of superconducting synchronous generators, *IEEE Tran. on Magnetics*, **MAG-19** (1983), pp. 522-8.
2 K. J. Binns and T. M. Wong, 'Analysis and performance of a high-field permanent-magnet synchronous machine', *IEE Proc. B, Electr. Power Appl.*, **131** (1984), pp. 252-8.
3 M. G. Say, *Alternating Current Machines*, 5th ed. (London: Pitman, 1983).
4 B. V. Jayawant, *Electromagnetic Levitation and Suspension Techniques* (London: Edward Arnold, 1981).
5 J. F. Eastham, 'Iron-cored linear synchronous machines', *Electronics and Power*, **23** (1977), pp. 239-42.
6 J. D. Edwards and A. M. El-Antably, 'Segmental-rotor linear reluctance motors with large airgaps', *Proc. IEE*, **125** (1978), pp. 209-14.
7 P. P. Acarnley, *Stepping Motors: a Guide to Modern Theory and Practice*, 2nd ed. (London: Peter Peregrinus, 1984).
8 W. F. Ray and R. M. Davis, 'Inverter drive for doubly salient reluctance motor', *IEE J. Electr. Power Appl.*, **2** (1979), pp. 185-93.
9 P. J. Lawrenson, J. M. Stephenson, P. T. Blenkinsop, J. Corda and N. N Fulton, 'Variable-speed switched reluctance motors', *IEE Proc. B, Electr. Power Appl.*, **127** (1980), pp. 253-65.

6 Induction Machines

6.1 Introduction

An essential feature of the operation of the synchronous machine is that the rotor runs at the same speed as the rotating magnetic field produced by the stator winding; the magnetic field as 'seen' from a point on the rotor does not vary with time. A very different type of machine results if the rotor is allowed to run more slowly than the rotating field; the rotor will 'see' a rotating field moving past it at the difference of the two speeds, and this field can induce currents in conductors on the rotor. The currents will interact with the rotating field to produce a torque, and this is the basic principle of the induction motor.

In common with other rotating machines, induction machines can operate as motors or generators. For reasons that will be discussed later, induction generators have a very limited use, and nearly all electric power is generated by synchronous machines. Induction motors, on the other hand, are used in far greater numbers than any other type of machine; they range in power rating from a few watts to tens of megawatts. The simplicity of the induction principle is reflected in the robust, reliable and relatively inexpensive construction of the machine itself, and the induction machine is the natural choice in the majority of motor applications.

The stator of an induction motor is similar to that of a synchronous motor, but the rotor structure is different. There are two forms of rotor: the cage rotor and the wound rotor. In a cage rotor, the conductors are in the form of bars (usually of aluminium or a copper alloy) which pass through slots in the laminated iron core of the rotor. The bars are connected to low-resistance rings (known as *end-rings*) at each end of the core, so that any pair of conductors forms a short-circuited turn. Cage rotors are cheap and robust, and in small sizes the bars and end-rings are die cast in a single operation. Figure 6.1 shows the construction of a small cage induction motor, and figure 6.2 shows the cage rotor of a large machine. It is possible to make the rotor in the same way as the stator, with a three-phase winding; an external resistor is connected to each phase of the rotor through sliprings, and the characteristics of the motor can be altered by varying the resistance. Figure 6.3 shows a typical rotor of this kind; it is much more expensive than the cage rotor, and is used only for special applications – see sections 6.4 and 6.5. Motors of this kind are known as *wound-rotor* or *slipring induction motors*. Wound rotors are also used in machines

Figure 6.1 Construction of a small cage induction motor (GEC Small Machines Ltd)

Figure 6.2 Cage rotor for a 21 MW 4-pole induction motor (GEC Large Machines Ltd)

Figure 6.3 Wound rotor for a 1.6 MW 8-pole slipring motor (GEC Large Machines Ltd)

known as *synchronous induction motors*; these operate as induction motors for starting, and are then made to run as synchronous motors by passing direct current through the rotor winding.

6.2 Electromagnetic action

The stator (or 'primary') winding of an induction machine is similar to that of a synchronous machine, and when connected to a suitable AC supply it will produce a rotating magnetic field of the form given by eqn (4.36)

$$B_1 = B_{1m} \cos(\omega t - \theta) \tag{6.1}$$

This equation implies that the machine has only two poles, and for simplicity the analysis in this section and in section 6.3 will be given for a two-pole machine. The results are extended at the beginning of section 6.4 to the general case of a $2p$-pole machine.

Suppose that the rotor of the induction machine rotates with an angular velocity ω_r, so that at time t a reference axis on the rotor makes an angle given by

$$\psi = \omega_r t \tag{6.2}$$

with the reference axis of the stator (figure 6.4). Let ϕ be the angular position

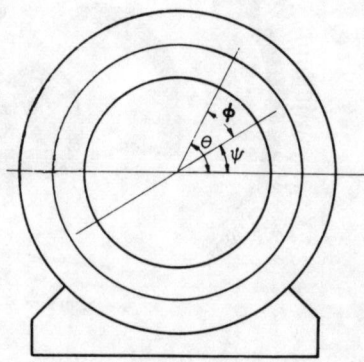

Figure 6.4 Angles and reference axes in the induction motor

of a point on the rotor, measured from the rotor reference axis; then the angle measured from the stator reference axis is

$$\theta = \phi + \psi = \phi + \omega_r t \tag{6.3}$$

Substitution of this expression for θ in eqn (6.1) gives the magnetic field in terms of the angle ϕ

$$B_1 = B_{1m} \cos\{(\omega - \omega_r)t - \phi\} \tag{6.4}$$

This equation implies that the rotor 'sees' a magnetic field rotating past it with an angular velocity of $\omega - \omega_r$, known as the *slip angular velocity*. The quantity

$$s = \frac{\omega - \omega_r}{\omega} \tag{6.5}$$

is termed the *fractional slip*, and we may rewrite eqn (6.4) in the form

$$B_1 = B_{1m} \cos(s\omega t - \phi) \tag{6.6}$$

This rotating field will induce an EMF at the slip angular frequency $s\omega$ in any conductor on the rotor.

Rotor magnetic field

Consider a rotor with a cage winding. An EMF at the slip angular frequency $s\omega$ will be induced in any conductor, and a current at this frequency will flow. The EMF is related to the flux density by eqn (4.54); consequently a pattern of currents will be set up in the rotor conductors which is similar in form to the magnetic field, and this pattern will rotate at the same speed $s\omega$. These rotor currents will in turn set up a rotating magnetic field of the form

$$B_2 = B_{2m} \cos(s\omega t - \phi - \delta_{12}) \tag{6.7}$$

The total magnetic field is thus $B = B_1 + B_2$, and this may be written as

$$B = B_m \cos(s\omega t - \phi - \delta_1) \tag{6.8}$$

It is this total field B which determines the EMF induced in a rotor conductor when rotor currents are flowing.

Rotating field concepts

The fields B_1 and B_2 rotate with the same angular velocity; with respect to the rotor, this is the slip angular velocity $s\omega$; with respect to the stator it is the synchronous speed ω. Since there is a constant angle δ_{12} between the axes of the fields, there will be a constant torque on the rotor given by

$$T = kB_{1m}B_{2m} \sin \delta_{12} \tag{4.30}$$

The mechanism of torque production is thus the same as in the synchronous machine, but the principle of operation is very different. In the synchronous machine, the rotor magnetic field is set up by externally impressed currents, and its axis is fixed relative to the rotor material. The rotor magnetic field of an induction machine, on the other hand, is produced by induced currents in the rotor, and its axis rotates relative to the rotor material. This rotation, or slipping, of the field past the rotor is an essential feature of induction motor operation; if the rotor ran at the synchronous speed it would 'see' a steady field, and there would be no induced currents.

6.3 Equivalent circuit

In order to predict the performance characteristics, it is necessary to construct an equivalent circuit which will represent the machine in terms of lumped circuit parameters and the stator terminal voltage and current. This may be done by considering the relationships between the magnetic field components, the currents flowing in the stator and rotor windings, and the EMFs induced in those windings. The fact that the rotor currents are induced from the stator makes the derivation of the equivalent circuit more difficult than the corresponding derivation for the synchronous machine.

Equivalent circuit of an ideal machine

Consider an induction machine with a cage rotor. The EMF induced in one rotor conductor may be calculated by the flux cutting rule in the same way as in eqn (4.54)

$$e_c = Blu = rls\omega B_m \cos(s\omega t - \phi - \delta_1) \tag{6.9}$$

If magnetic leakage is neglected, then a current proportional to this EMF will flow in the conductor. Thus a pattern of induced currents will be set up in the rotor with a distribution which matches the sinusoidal distribution of the magnetic field.

Equation (6.9) shows that there will be no induced EMF, and therefore no current, when $s = 0$; the rotor speed ω_r is then equal to the rotating field speed ω. Figure 6.5 shows the stator current density and the airgap flux density under these conditions, while figures 6.6 and 6.7 show the corresponding space and time phasor diagrams. It is assumed for the present that there is negligible stator resistance or leakage reactance, so that the stator terminal voltage V_1 is equal to the EMF induced by the rotating field.

Suppose that the rotor speed ω_r is allowed to fall below the synchronous speed ω, but the stator terminal voltage V_1 is held constant. Current will flow in the rotor, as shown in figure 6.8. Since the stator voltage is constant, the total

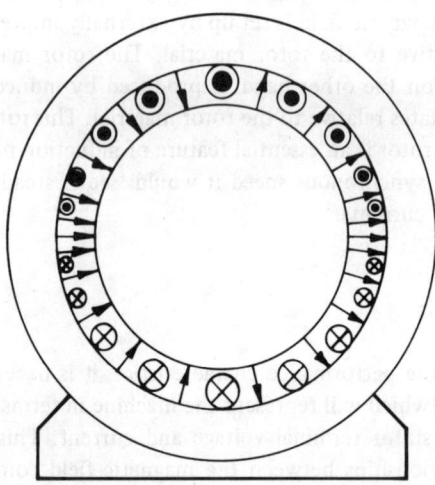

Figure 6.5 Current–density and flux–density distributions for $\omega_r = \omega_s$

INDUCTION MACHINES 163

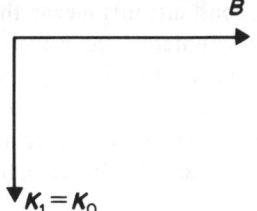

Figure 6.6 Space phasor diagram for $\omega_r = \omega_s$

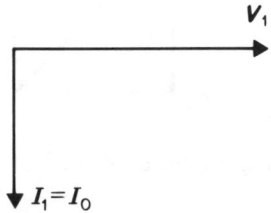

Figure 6.7 Time phasor diagram for $\omega_r = \omega_s$

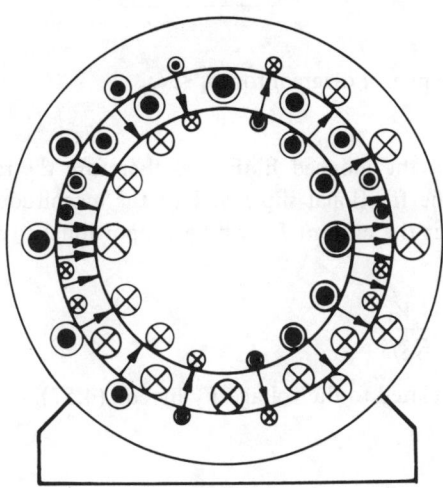

Figure 6.8 Current-density and flux-density distributions for $\omega_r < \omega_s$

flux density B must remain constant; this means that the stator current density K_1 must change in order to counteract the effect of the rotor current density K_2. We may consider K_1 to be made up of two parts, as shown in figures 6.8 and 6.9: a component K_2' which is equal and opposite to the rotor current density K_2, and a component K_0 which sets up the magnetic field. There will be corresponding current components I_2' and I_0 (figure 6.10), which make up the total stator phase current I_1.

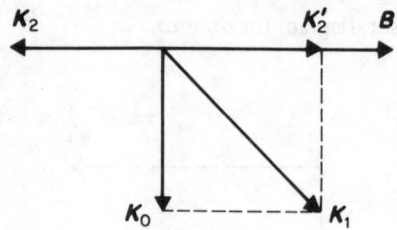

Figure 6.9 Space phasor diagram for $\omega_r < \omega_s$

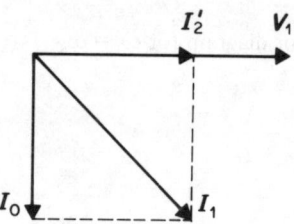

Figure 6.10 Time phasor diagram for $\omega_r < \omega_s$

From eqn (6.9) the induced EMF, and therefore the rotor current, will be proportional to the fractional slip s and to the magnitude of the flux density B_m. It follows that the current I_2' will be proportional to s and to V_1, and we may put

$$I_2' = \frac{sV_1}{R_2'} = \frac{V_1}{R_2'/s} \tag{6.10}$$

The current I_0 is related to the voltage V_1 by eqn (4.57), so that

$$I_0 = \frac{V_1}{jX_m} \tag{6.11}$$

Equations (6.10) and (6.11) are the equations of the circuit shown in figure 6.11, which is the equivalent circuit for an ideal induction machine. The element

INDUCTION MACHINES 165

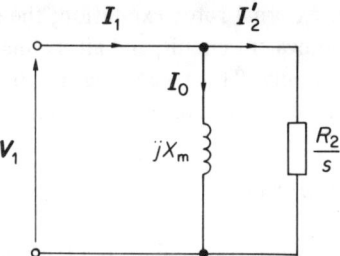

Figure 6.11 Equivalent circuit of an ideal induction machine

R'_2/s represents the effects of the rotor (or secondary) in the stator (or primary) circuit.

Complete equivalent circuit

The similarity of the induction motor to a transformer with a closed secondary circuit should be noted; the current I'_2 is the secondary (rotor) current referred to the primary. The relative motion between the primary and secondary is represented in the equivalent circuit by the factor $1/s$ multiplying the secondary resistance R'_2, and the significance of this term will be explained later. By analogy with the transformer, the effects of stator resistance, stator and rotor leakage reactance and core loss may be included in the equivalent circuit. Figure 6.12 shows the complete equivalent circuit of the machine obtained in this way. For an alternative derivation, see section 7.5.

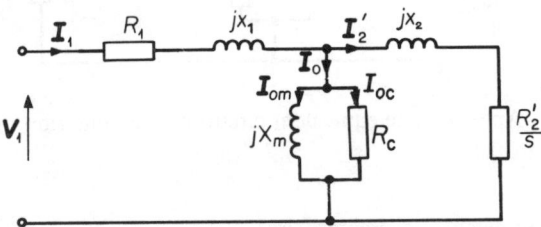

Figure 6.12 Complete equivalent circuit of the induction machine

It is instructive to compare the equivalent circuit of the induction machine with the complete equivalent circuit of the synchronous machine shown in figure 5.8. In the synchronous-machine equivalent circuit, the rotor is represented by an active element — a current generator — which is capable of supplying some or all of the magnetising current I_{om}. In the induction machine, on the other hand, there is no external rotor excitation; the rotor is represented by a passive element in the equivalent circuit, and all the magnetising current must be drawn from the stator supply. This means that an induction motor necessarily behaves as an inductive load, taking current at a lagging power factor.

Approximate equivalent circuit

The equivalent circuit shown in figure 6.12 is a fairly accurate representation of the machine, and it may be used to predict the characteristics. The analysis is greatly simplified, however, if two approximations are made. First the shunt elements R_c and X_m are transferred to the input terminals. This was shown to be a good approximation with a power transformer (see section 3.4); it is less satisfactory with the induction machine, because the airgap between the stator and rotor reduces the value of X_m and increases x_1, in comparison with a transformer of similar rating. The second approximation is to ignore the stator resistance R_1 in comparison with the term R'_2/s. This is a good approximation when the slip is small, and the equivalent circuit then takes the form shown in figure 6.13; the primary and secondary leakage reactances are combined to give a total leakage reactance $X = x_1 + x'_2$. This simplified circuit demonstrates the essential features of induction motor performance, and the complete equivalent circuit can always be used when a more accurate calculation is required.

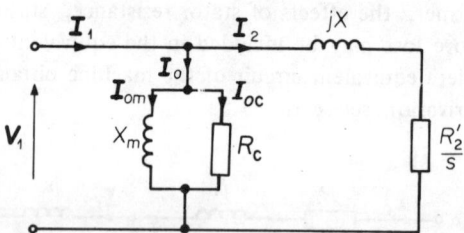

Figure 6.13 Approximate equivalent circuit of the induction machine

6.4 Induction machine characteristics

Multi-pole machines

If the machine stator winding has p pairs of poles, the synchronous speed is $\omega_s = \omega/p$. The fractional slip is now defined as

$$s = \frac{\omega_s - \omega_r}{\omega_s} \tag{6.12}$$

and the slip angular velocity is $\omega_s - \omega_r = s\omega_s$. Since the rotor currents also generate p pairs of poles, the angular frequency of the rotor currents will be p times the slip angular velocity: $ps\omega_s = s\omega$. Thus for a given slip, the rotor frequency is unchanged. The basic electromagnetic action is also unchanged, and the same equivalent circuit holds for a machine with any number of pole pairs.

Rotor power relationships

In the equivalent circuit of figure 6.12, the power loss in R_1 and R_c represents the primary copper loss and the core loss. The power loss in the resistance R'_2/s must therefore represent the average input of power to the rotor, for there can be no dissipation of power in the reactances X_m and x'_2. Thus the input of power per phase to the rotor is $(I'_2)^2 R'_2/s$; but the power dissipated in the actual resistance of the rotor circuit is only $(I'_2)^2 R'_2$. The difference between these quantities is

$$(I'_2)^2 R'_2 \frac{1-s}{s}$$

and this must represent electrical energy converted into mechanical form. If P is the total power absorbed by the rotor, then

$$\text{electromagnetic power input to rotor} = m(I'_2)^2 \frac{R'_2}{s} = P \qquad (6.13)$$

$$\text{power loss in rotor resistance} = m(I'_2)^2 R'_2 = sP \qquad (6.14)$$

$$\text{mechanical power output} = m(I'_2)^2 R'_2 \frac{1-s}{s} = (1-s)P \qquad (6.15)$$

where m is the number of phases.

Consider the torque T exerted on the rotor by the rotating magnetic field. If there are p pairs of poles, the angular velocity of the field is $\omega_s = \omega/p$, and the rotating field therefore does work at the rate $\omega_s T$. This would obviously be true if the rotating field were produced by physical poles on the stator, driven mechanically at a speed ω_s; the electromagnetic field is the same when a polyphase winding produces the rotating field, so the work done must be the same. Since the rotor runs at a speed $\omega_r = (1-s)\omega_s$, the mechanical power output is $\omega_r T = (1-s)\omega_s T$. The difference between the work done by the field and the mechanical output must be absorbed in rotor losses, so this is $(\omega_s - \omega_r)T = s\omega_s T$. Thus

$$\text{electromagnetic power input to rotor} = \omega_s T \qquad (6.16)$$

$$\text{power loss in rotor resistance} = (\omega_s - \omega_r)T = s\omega_s T \qquad (6.17)$$

$$\text{mechanical power output} = \omega_r T = (1-s)\omega_s T \qquad (6.18)$$

Note that the fraction of the input power lost in rotor resistance is equal to the fractional slip s; since there must always be some slip between the rotor and the rotating magnetic field, this represents an unavoidable power loss. The ratio of mechanical power output to electromagnetic power input is termed the *rotor efficiency*, and its value is $1 - s$.

Torque/speed characteristics

The torque may be calculated for a given value of slip by equating expressions (6.13) and (6.16) for electromagnetic power, and obtaining the value of I_2' from the equivalent circuit of figure 6.13. For a machine with m phases and p pole pairs, the result is

$$T = \frac{mp}{\omega} \cdot \frac{R_2'}{s} \cdot \frac{V_1^2}{X^2 + (R_2'/s)^2} \tag{6.19}$$

which may be written in the alternative form

$$T = \frac{mp}{\omega} \cdot \frac{V_1^2}{X} \cdot \frac{1}{sX/R_2' + R_2'/sX} \tag{6.20}$$

Since the slip s is related to the rotor speed ω_r by eqn (6.12), eqn (6.20) gives the torque/speed relationship for the induction machine. Figure 6.14 shows a typical torque/speed or torque/slip characteristic. Note that there can be no induced rotor currents when $s = 0$ and $\omega_r = \omega_s$, so the torque must be zero at this point. There are three distinct regions to the torque/speed characteristic shown in figure 6.14, which will be considered in turn.

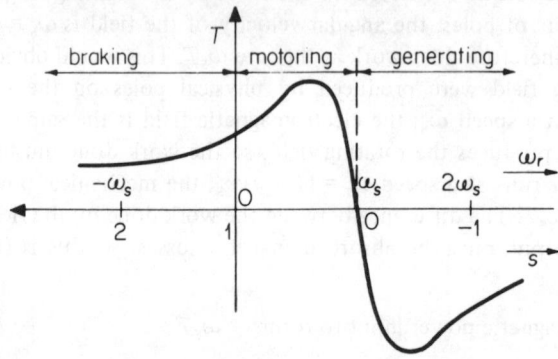

Figure 6.14   Torque/speed characteristic for the induction machine

Motoring region

In this region the rotor speed ω_r is positive but less than the synchronous speed ω_s; the torque is also positive, and the machine converts electrical power into mechanical power. The value of the slip s varies from 1 when the rotor is stationary to 0 when the rotor runs at the synchronous speed. In the complete equivalent circuit (figure 6.12) the resistance R'_2/s is positive and greater than R'_2; thus the total electrical power absorbed by the rotor exceeds the power dissipated in the rotor resistance, and the balance is extracted as mechanical power at the shaft. Since the overall efficiency of the machine cannot exceed the rotor efficiency of $1 - s$, induction motors normally operate with a small value of slip. The full-load slip can be as low as 1 per cent in large machines and seldom exceeds 5 per cent in small machines, so the normal rotor speed is always close to the synchronous speed.

Generating region

When the rotor is driven mechanically so that its speed exceeds the synchronous speed, the torque reverses and the machine absorbs mechanical power. The slip is negative in this region, and the resistance R'_2/s is also negative; the rotor therefore exports electrical power to the stator, and the machine acts as a generator. The machine must normally be connected to an AC supply before it will act as a generator, since a source of reactive power is required for the magnetising current flowing in X_m which cannot be provided by the rotor. The synchronous machine provides its own magnetising current by having a field winding on the rotor, and this is the preferred type of AC generator. An induction generator is sometimes used when it is not possible to obtain the constant-speed drive required by a synchronous generator, for example in a wind-powered generator.

Braking region

If the rotor is driven in the opposite direction to the rotating field, the torque will oppose the motion and the machine will again absorb mechanical power. It will not, however, act as a generator. The slip is now greater than 1, and the resistance R'_2/s is positive and smaller than R'_2; the electrical power input to the rotor is less than the rotor resistance loss, and the balance is supplied by the mechanical power input. The machine therefore acts as a brake, with both the electrical and the mechanical power inputs dissipated in the rotor resistance.

Operation in the braking region can only take place for short periods on account of rotor heating; it is sometimes used as a method of rapidly stopping an induction motor by 'plugging', as follows. Two of the connections to the three-phase stator are interchanged, thus reversing the direction of rotation of the magnetic field and applying a braking torque to the rotor. The stator supply must be disconnected as soon as the rotor comes to rest, otherwise the rotor will continue to accelerate in the new direction of the rotating field. Another method of braking is to pass direct current through the stator winding: see problem 6.3 and section 7.5.

Induction motor torque

Several interesting properties of the torque characteristic may be deduced from eqns (6.19) and (6.20). When the slip is such that $sX/R'_2 = 1$, the torque will have a maximum value given by

$$T_m = \frac{mpV_1^2}{2\omega X} \qquad (6.21)$$

This is known as the *breakdown torque*, and a load in excess of this value will stop the motor. The normal operating region lies to the right of the peak, and the normal full-load torque is usually less than half the breakdown torque. If s_m denotes the value of slip corresponding to the maximum torque, then

$$\frac{T}{T_m} = \frac{2}{s/s_m + s_m/s} \qquad (6.22)$$

The value of the breakdown torque is determined by the total leakage reactance X, and the slip at which it occurs is determined by the rotor resistance R'_2. Figure 6.15 shows a family of torque/speed curves for different values of R'_2, and it will be seen that increasing the rotor resistance increases the standstill

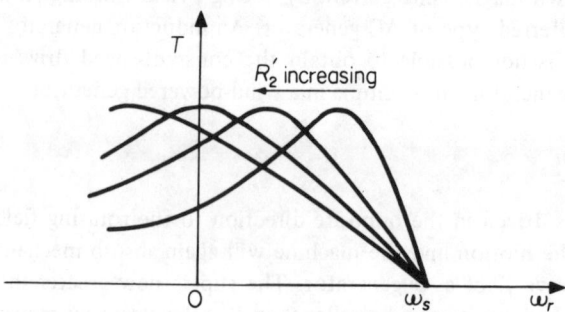

Figure 6.15 Torque/speed characteristics for varying R'_2

torque (the torque when the rotor is stationary). Unfortunately this also detracts from the full-load performance, for the following reason. At small values of slip, R'_2/s is very much greater than X, and eqn (6.19) becomes

$$T = \frac{mpV_1^2}{\omega} \cdot \frac{s}{R'_2} \qquad (6.23)$$

Thus the torque/slip characteristic is linear in this region, with a slope inversely proportional to R'_2. Equation (6.23) also shows that s must increase in proportion to R'_2 if the machine is to develop the same full-load torque. The choice of R'_2 therefore involves a compromise between the starting torque and the full-load slip, which in turn affects the efficiency and the speed regulation.

Current, power factor and efficiency

The form of the equivalent circuit (figure 6.12) shows that each phase of an induction motor will act as an inductive impedance, and the machine therefore takes current at a lagging power factor. Since the rotor impedance $R'_2/s + jx'_2$ varies in magnitude and phase angle with the slip s, the stator current I_1 will also vary with slip (and hence with the rotor speed ω_r). Figures 6.16(b) and 6.16(c) show the variation of the stator current magnitude I_1 and the power factor $\cos \phi$ (where ϕ is the phase angle between V_1 and I_1) for a typical small cage induction motor. The full-load speed is shown in the figure, and it will be seen that the starting current when the rotor is stationary is about 5 times as large as the full-load running current. Induction motors are frequently started simply by connecting the stator directly to the supply (direct-on-line or DOL starting). When the resulting high starting current is unacceptable, the phase voltage can be reduced either by the use of auto-transformers or series reactors, or by using star connection of the windings for starting and changing over to delta connection after the rotor has accelerated. For small motors, power electronic controllers can provide a 'soft start' by gradually increasing the applied voltage from zero up to the full mains value.

The efficiency of an induction motor is defined in the usual way as the ratio of the mechanical output power to the electrical input power. A typical efficiency curve is shown in figure 6.16(d), and the flow of power through the machine may be traced in the same way as for the DC motor (section 2.3); for an m-phase machine we have

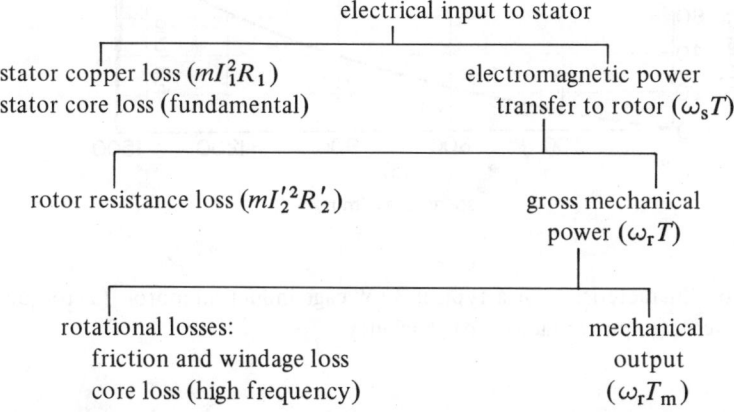

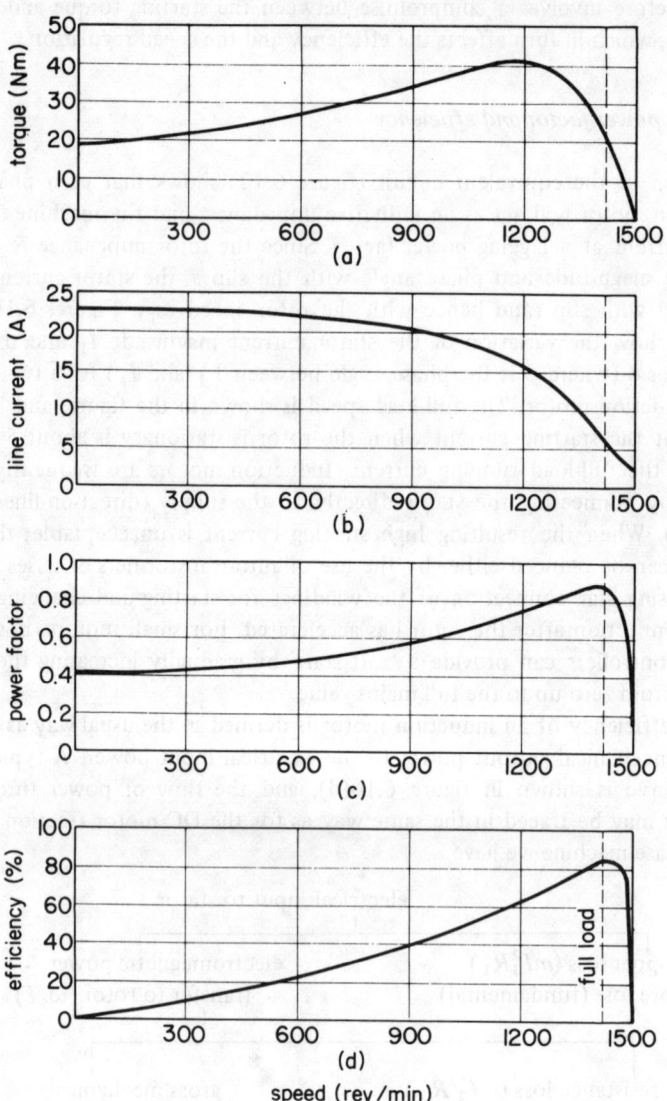

Figure 6.16 Characteristics of a typical 3 kW cage induction motor: (a) torque, (b) line current, (c) power factor, (d) efficiency

The important difference between this flow diagram and the corresponding diagram for the DC machine lies in the separation of the core loss into two components. The fundamental core loss, due to the fundamental component of the rotating magnetic field, is mainly confined to the stator core. This follows from the fact that the magnetic field at a point in the stator alternates with the supply angular frequency ω, whereas the corresponding field at a point fixed in the rotor alternates with the slip angular frequency $s\omega$; fundamental eddy-current and hysteresis losses are therefore insignificant in the rotor at the normal full-load slip. There is another component of core loss caused by (a) harmonic components of the rotating field which arise from the non-sinusoidal distribution of practical windings, and (b) pulsations in the field which arise from the relative motion of rotor and stator slots. This high-frequency component of core loss occurs in both the rotor and the stator, and the energy is supplied in a very complex way; it is customary to assume that it can be represented by a rotational loss term added to the mechanical (windage and friction) losses.

Figure 6.16(c) and (d) show that the power factor and efficiency can be low when the motor is operating at less than full load from a constant supply voltage. Power electronic controllers are now available [1] which can adjust the supply voltage automatically so that the motor operates at optimum power factor under all load conditions.

6.5 Speed control of induction motors

Induction motors account for about 90 per cent of the electrical drives used in industry, and the majority of these applications require a fairly constant speed. Induction machines are normally designed to work with a small value of slip (generally less than 5 per cent) at full load, and the deviation of the rotor speed ω_r from the synchronous speed ω_s is therefore small. There are certain applications, however, which require substantial variation of the motor speed. DC motors form an obvious choice for this kind of drive because of the ease of speed control, but they are relatively expensive. The induction motor has the advantages of low cost and high reliability, and the possibility of controlling its speed is now examined.

The possible methods of speed control may be deduced from eqn (6.12), which defines the fractional slip s. Thus

$$\omega_r = (1-s)\omega_s = (1-s)\frac{\omega}{p} \tag{6.24}$$

showing that the rotor speed may be controlled by varying the slip s, the number of pole-pairs p, or the supply angular frequency ω. These methods will be considered in turn.

Variation of rotor slip

For a given load torque T, eqn (6.20) shows that the quantity sX/R'_2 is a constant. Increasing the resistance R'_2 will cause a proportionate increase in the slip s, with a consequent decrease in the rotor speed. In practice the load torque will vary with the speed, and the precise effect of varying R'_2 may be found by plotting the torque/speed characteristic for the load on the graph of motor torque/speed characteristics shown in figure 6.15. The point of intersection of the load curve with the motor characteristic gives the speed for each value of R'_2, as shown in figure 6.17. Because of the power loss associated with the slip, this is an inefficient method of speed control. It is often used for short periods when a large starting torque is required; a slipring motor is employed, and the external resistance is reduced to zero as the rotor runs up to speed. When continuous operation at high slip is required for speed control purposes, the slip power can be extracted from the rotor circuit and returned to the mains via a frequency converter; this is the basis of the slip power recovery or Kramer system [2]. An alternative method of varying the slip, which may be applied to cage rotor machines, is to vary the magnitude of the stator voltage V_1. Figure 6.18 shows the family of torque/speed characteristics for a number of values of V_1, and the points of intersection with the load curve give the corresponding values of speed. As with rotor resistance variation, this is an inefficient method of speed control. It has the merit of simplicity, and it is sometimes used with small machines when efficiency is not particularly important.

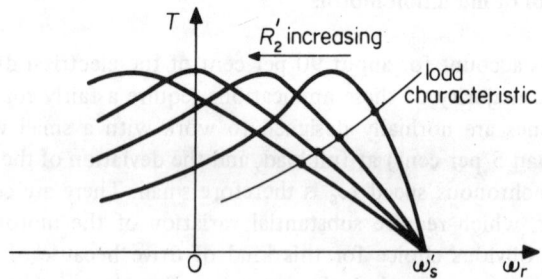

Figure 6.17 Speed control by variation of rotor resistance

Pole-change windings

The second method of speed control is by alteration of the number of pole-pairs p; this can only give discrete changes of speed, since p must be an integer. With a properly designed cage rotor it is only necessary to alter the number of poles of the stator winding, for the corresponding rotor currents will find their own paths in the cage. An obvious way of varying p is to have an independent winding

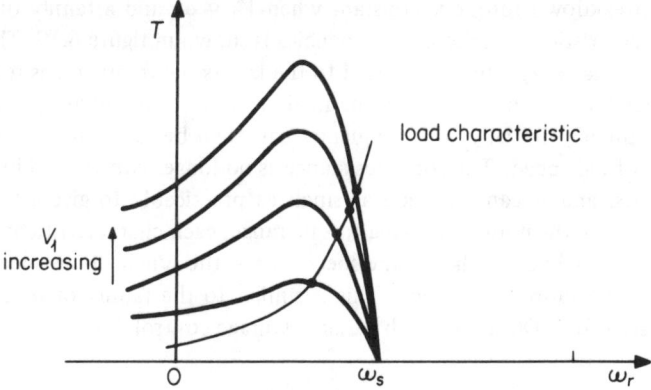

Figure 6.18 Speed control by variation of supply voltage

for each pole number, with a selector switch to connect the appropriate winding to the supply. A better solution is to design a single winding in such a way that the number of poles can be changed merely by altering the interconnection of the coils. The technique of pole-amplitude modulation [3] permits values of p such as 4, 5, 6 to be obtained from a single stator winding, and this gives a useful degree of speed control.

Frequency variation

The third and most interesting method of speed control is achieved by variation of the supply angular frequency ω. This permits continuous variation of the speed; the slip can be kept low to maintain the efficiency; and the method can be applied to cage induction motors. It is desirable to maintain a constant flux density in the airgap when the frequency is varied, and eqn (4.55) shows that the magnitude of the supply voltage must vary in proportion to the frequency if B_m is to remain constant. This may also be deduced from the approximate equivalent circuit (figure 6.13), for the magnetising current is given by

$$I_{0m} = \frac{V_1}{X_m} = \frac{V_1}{\omega L_m} \tag{6.25}$$

and this will be constant if $V_1 \propto \omega$. The torque is given by eqn (6.20), and if the reactance X is written as ωL, this becomes

$$T = \frac{mp}{L} \cdot \frac{V_1^2}{\omega^2} \cdot \frac{1}{s\omega L/R_2' + R_2'/s\omega L} \tag{6.26}$$

Thus the breakdown torque is constant when $V_1 \propto \omega$, and a family of torque/speed characteristics for different frequencies is shown in figure 6.19. The curves all have a similar shape, but are shifted to the left as the frequency is reduced. It follows that the starting torque can be made as large as required by using a low enough frequency initially; the frequency can then be raised to accelerate the rotor to its final speed. The rotor resistance is no longer constrained by starting requirements, and it can be made as small as practicable to give a low slip at normal load. For the normal working torque range, each characteristic is approximately a straight line which cuts the speed axis at the synchronous speed corresponding to the supply frequency. This is similar to the family of torque/speed characteristics for a DC motor with variable-voltage control.

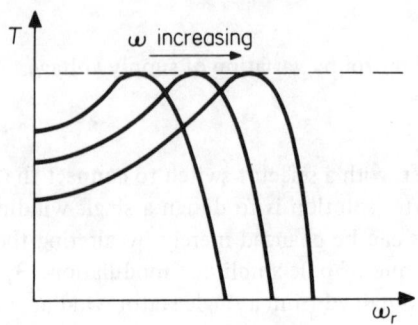

Figure 6.19 Torque/speed characteristics for varying ω

The variable-frequency stator supply may be obtained from the fixed-frequency AC mains via a solid-state frequency converter or inverter [4]. Transistor inverters are normally used with small motors, employing pulse-width modulation (PWM) to simulate a sinusoidal supply waveform. Thyristor inverters are normally used with large motors, and they usually generate a quasi-square waveform instead of sine-wave PWM. At low frequencies the stator resistance R_1 cannot be neglected, and the supply voltage V_1 must be increased to maintain constant flux; this is done automatically by the inverter control circuit.

Frequency-controlled induction motors are now comparable in cost with voltage-controlled DC motors for variable-speed drives; the higher cost of the electronic controller is offset by the lower cost of the motor. Induction motors require little maintenance, and are better suited than DC motors to operation in hazardous or dusty environments. As with DC drives, operation from an electronic controller can cause additional losses in the induction motor. With sine-wave PWM inverters the effect is negligible; but a quasi-square waveform can cause an increase of 10-20 per cent in the motor loss [5], so the motor must be derated to avoid overheating.

6.6 Single-phase induction motors

Large induction motors have three-phase stator windings, and many small industrial induction motors are also made this way. Light industrial and domestic applications require a different kind of induction motor which can operate from a single-phase supply [6, 7]. One approach is to use a two-phase machine, with a capacitor connected in series with one phase of the stator winding (figure 6.20). The current I_α in the phase connected directly to the supply will lag the supply voltage V by an angle α (figure 6.21). By a suitable choice of the capacitance value, the current I_β in the other phase can be made to lead the voltage V by an angle β. If $\alpha + \beta = \pi/2$, the currents are in quadrature, and the motor will operate as a normal two-phase machine. Since the impedance presented by each phase of the stator winding varies with the load on the motor, the phase splitting will only be exact for one particular value of load torque, and there will be some unbalance between the phases at other load conditions. Machines of this kind are known as capacitor motors.

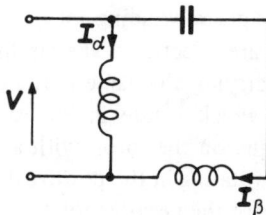

Figure 6.20 Single-phase capacitor motor

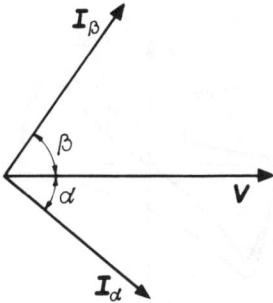

Figure 6.21 Phasor diagram for the capacitor motor

An alternative approach is to take a machine with a normal cage rotor, but only a single stator winding connected to the single-phase supply. If a current i flows in this winding, the current density is given by eqn (4.12)

$$K = -Zi \sin \theta$$

Since i is a sinusoidal alternating quantity of the form

$$i = I_m \cos \omega t$$

the current density is

$$K = -ZI_m \cos \omega t \sin \theta \tag{6.27}$$

By a trigonometric identity this may be written in the form

$$K = -ZI_m \tfrac{1}{2}\{\sin(\omega t - \theta) + \sin(\omega t + \theta)\} \tag{6.28}$$
$$\quad\quad\quad\quad\quad\quad (a) \quad\quad\quad\quad (b)$$

Term (a) in eqn (6.28) represents a field rotating with angular velocity ω in the positive direction; while term (b) represents a field rotating with angular velocity ω in the negative direction. Thus the pulsating field produced by alternating current flowing in a single-phase winding may be resolved into two rotating fields which rotate in opposite directions. The machine behaves as though it had two polyphase windings, carrying the same current magnitude per phase, but producing magnetic fields which rotate in opposite directions. Each rotating field will give rise to a torque on the rotor, with a corresponding torque/speed characteristic. For the field rotating in the positive direction, this is similar to the normal torque/speed curve; for the negative rotating field the direction of torque is reversed, and zero slip now corresponds to a rotor speed of $-\omega_s$. Figure 6.22 shows the two torque/speed curves, together with the resultant torque which is the sum of the two components. The torque characteristic for each component field is somewhat different from the normal torque/speed characteristic, because the EMF induced by the other field component affects the current which the stator draws from the supply. In consequence the braking torque is reduced, and

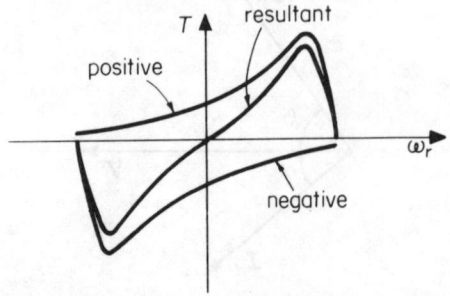

Figure 6.22 Torque/speed characteristic for a single-winding single-phase motor

the net output torque is greater than would be given by the normal torque/speed characteristics: see problem 6.6.

The resultant torque is zero when the rotor is stationary, so the machine is not inherently self-starting. If some means is provided for spinning the rotor, the resultant torque acts in a direction to accelerate the rotor, which will run up to speed in the normal way. The starting torque is provided by an auxiliary winding in space quadrature with the main winding; the field produced by this winding combines with a portion of the main field to produce a rotating field, which exerts a torque on the stationary rotor. There are three common arrangements for the auxiliary winding.

Shaded-pole motor

This simple form of single-phase motor is shown in figure 6.23. The stator has salient poles carrying the main winding, and a portion of each pole is enclosed by a ring which is usually made of copper. Currents are induced in the ring (or 'shading coil') by the alternating magnetic field, and the portion of the pole enclosed by the ring is 'shaded' from the main pole flux; the flux is weaker, and its phase is retarded relative to the main flux. The arrangement forms a rudimentary two-phase winding which is adequate for accelerating the rotor against light loads. Motors of this kind are robust and inexpensive; but they are also inefficient, and their use is restricted to sizes below about 250 W.

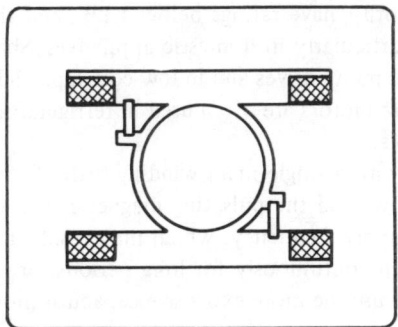

Figure 6.23 Shaded-pole induction motor

Split-phase motor

For larger sizes of single-phase induction motor (up to about 1 kW) with modest starting torque requirements, a normal stator construction is used. The auxiliary winding is designed to have very different values of resistance and reactance from the main winding, so that there is an appreciable phase shift between the currents when the two windings are connected to the same single-phase supply. Thus the phase-splitting is inherent in the machine design, and the auxiliary winding normally has a short time rating. It is disconnected once the rotor has run up to speed, either by a centrifugally operated switch or by a relay which senses when the main winding current has fallen from a high starting value to the normal running value.

Capacitor-start motor

When the motor must develop a large starting torque, the capacitor motor arrangement is used. This gives a larger phase angle between the main and auxiliary currents (ideally 90°), and permits a better design of the auxiliary winding since the phase shift is provided by the capacitor. As with the split-phase motor, the auxiliary winding is disconnected once the rotor has run up to speed. The capacitor is usually an AC electrolytic type with a short time rating; this is much cheaper than the type of capacitor required for continuous operation in a capacitor motor.

Applications

Single-phase induction motors have lower values of efficiency and power factor than comparable polyphase machines, and their use is restricted to powers below about 4 kW. The majority have ratings below 1 kW, and they are employed in very large numbers, particularly in domestic appliances. Shaded-pole motors are used for small fan and pump drives and in low-cost tape and record decks. Split-phase or capacitor-start motors are often used in refrigerators, washing machines and small machine tools.

Motors which run with a single main winding suffer from two disadvantages: the power factor is low; and the pulsating magnetic field causes the torque to pulsate at twice the supply frequency, which may result in noise and vibration. If the motor has to run continuously for long periods, or if noise is a problem, it may be preferable to use the more expensive capacitor motor.

6.7 Linear induction motors

The induction-motor principle can be made to produce linear motion by the simple expedient of placing the stator coils in a flat iron core instead of a cylindrical core. Three-phase currents flowing in the stator, or primary, coils will produce a current-density pattern which travels along the surface; this will set up a travelling magnetic field analogous to the rotating magnetic field in a rotary induction motor. If a flat conducting plate is placed next to the primary, the field will induce currents in the plate, and the interaction of these currents with the field will produce a force tending to move the plate in the direction of motion of the field. The plate usually has an iron backing to complete the magnetic circuit, and it is the linear counterpart of the rotor of a conventional motor; it is termed the *secondary*. If the secondary were made the same length as the primary, then the two would soon separate; consequently linear induction motors are made in two forms: short-primary machines and short-secondary machines (figure 6.24). These are the most common forms of linear induction motor, but several other arrangements are possible [7, 8, 9].

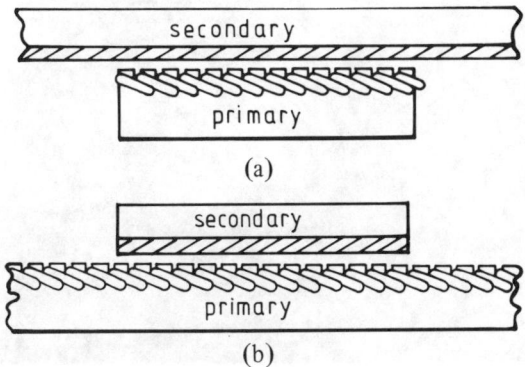

Figure 6.24 Linear induction motors: (a) short primary; (b) short secondary

Linear induction motors differ from rotary machines in two important respects. Firstly, the airgap is usually much larger; this entails a large magnetising current, and in consequence the power factor and efficiency are low. Secondly, the magnetic field decays at the ends of the primary; in a rotary machine, the rotating field closes on itself and is effectively endless. In a short-primary machine, currents in the secondary are largely confined to the region of the primary; currents have to build up as secondary material approaches the entry edge, and then decay as material leaves the exit edge. These transient currents set up field components travelling at different speeds from the primary current wave, and they modify the performance of the machine in a complex way [3, 8, 10]. The effects depend on the slip; they result in a reduction of the force on the secondary and an increase in the losses.

Applications of linear induction motors are numerous [7, 9]. Industrial uses include the propulsion of overhead travelling cranes and the handling of sheet metals such as aluminium. They are widely proposed for advanced ground transport systems [11]; several prototype systems have been demonstrated, and figure 6.25 shows the stator of a linear motor for a low-speed passenger vehicle which is now in commercial service in Birmingham [12].

Figure 6.25 Stator of a linear induction motor for a passenger vehicle (Brush Electrical Machines Ltd)

Problems

6.1. In problem 2.3, the DC shunt motor is replaced by a slipring induction motor with an external resistance R in each phase of the rotor circuit. If the internal losses of the motor may be neglected, show that the same expressions hold for the hoisting speed and the efficiency.

6.2. If the stator of an induction motor is supplied with a constant current I_1, and the rotor leakage reactance may be neglected, obtain an expression for the torque in a form similar to eqn (6.20). If s_m is the slip for maximum torque, compare the values of s_m for constant-current and constant-voltage operation.

6.3. A cage induction motor has direct current applied to the stator winding. Show that the machine acts as a brake for both directions of rotation, and deduce the shape of the torque/speed curve by considering the slip speed of an induction motor operating from a constant-current AC supply.

6.4. When the rotor of an induction motor accelerates from rest with no mechanical load coupled to the shaft, part of the energy input to the rotor will be dissipated as heat in the rotor resistance and part will be stored as kinetic energy of the rotor. If rotational losses may be neglected, show that when the rotor reaches the synchronous speed the total energy dissipated in the rotor resistance is equal to the final kinetic energy of the rotor.

6.5. An induction motor has a two-phase stator winding, and it runs with a slip s when positive-sequence voltages V_p and $-jV_p$ are applied to the α and β phases respectively. With the rotor speed unchanged the positive-sequence supply is disconnected and negative-sequence voltages V_n, jV_n are applied to the α and β phases respectively. Show that the slip is now equal to $2-s$, and draw complete equivalent circuits for the two conditions of operation.

6.6. In problem 6.5, the voltages V_p and V_n are adjusted so that in the first case the α and β phase currents are I, $-jI$ and in the second they are I, jI. Use the principle of superposition to show that the motor operates as a single-phase machine when voltages $(V_p + V_n)$, $(-jV_p + jV_n)$ are applied to the α and β phases respectively, and hence combine the two separate equivalent circuits into a single equivalent circuit for the single-phase machine. Note that the currents in a circuit will be doubled if all the impedance values are halved.

References

1. L. Holmes, 'Energy-saving motor controllers', *Electronics and Power*, 28 (1982), pp. 232-5.
2. A. E. Fitzgerald, C. Kingsley Jr. and S. D. Umans, *Electric Machinery*, 4th ed. (New York: McGraw-Hill, 1983).
3. B. V. Jayawant, *Induction Machines* (Maidenhead: McGraw-Hill, 1968).
4. J. Hindmarsh, *Electrical Machines and their Applications*, 4th ed. (Oxford: Pergamon Press, 1984).
5. J. M. D. Murphy, *Thyristor Control of AC Motors* (Oxford: Pergamon Press, 1973).
6. P. L. Alger, *Induction Machines*, 2nd ed. (New York: Gordon and Breach, 1970).
7. M. G. Say, *Alternating Current Machines*, 5th ed. (London: Pitman, 1983).
8. E. R. Laithwaite, *Induction Machines for Special Purposes* (London: Newnes, 1966).
9. E. R. Laithwaite, *Linear Electric Motors* (London: Mills and Boon, 1971).
10. S. Yamamura, *Theory of Linear Induction Motors*, 2nd ed. (University of Tokyo Press, 1978).
11. B. V. Jayawant, *Electromagnetic Suspension and Levitation Techniques* (London: Edward Arnold, 1981).
12. H. Linacre, J. S. Chahal, G. Crawshaw and B. Rawlinson, 'Birmingham Airport maglev propulsion system', *IMechE International Conference on Maglev Transport, October 1984, C408/84*, pp. 193-201.

7 Generalised Machine Theory

7.1 Introduction

We have seen that there is a close relationship between the induction machine and the synchronous machine. The brushless DC motor mentioned in section 2.7 was shown to be a form of synchronous machine, and the same is true of the conventional DC machine. In a normal synchronous machine, the field poles rotate and the AC armature winding is stationary. This structure can be inverted, so that the field poles are on the stator and the AC armature is on the rotor; this is similar to the structure of a DC machine, and small synchronous machines are often made in this form. The similarity goes further than this; in a DC machine armature, the coil currents and voltages are alternating quantities, just as they are in the synchronous machine. We may therefore regard the DC machine as a special form of synchronous machine; the commutator converts DC at the terminals into AC in the coils, and vice versa, and it also constrains the magnetic fields of the stator and rotor to be at right angles. In principle a synchronous motor could be made to behave like a DC motor if the AC supply frequency were controlled so as to make the angle δ_{12} (figure 5.2) 90° at all times. This essential unity of the different types of rotating machine suggests the possibility of a unified mathematical theory. Such a theory was developed by Kron in the 1930s, using the mathematical methods of tensor calculus [1, 2]. It is possible to express this theory in terms of matrices, and a key element is a mathematical transformation of variables which is equivalent to the physical action of the commutator. This transformation enables the basic equations of AC and DC machines to be written in identical form, and a routine process permits these equations to be set up by inspection.

Two forms of model, or 'primitive', machine are used in the generalised theory: a slipring model and a commutator model. The slipring model is derived from the inverted synchronous machine, and has polyphase windings on the stator and the rotor; for simplicity these are two-phase windings, and for generality the stator can have salient poles (figure 7.1). Connection to the rotor winding is made via sliprings. In the commutator model the stator is the same, but the rotor carries a commutator winding with two sets of brushes (figure 7.2). There is no loss of generality in choosing two-phase windings; it is shown in appendix B that for any arbitrary three-phase currents flowing in a three-phase winding, we can always find the corresponding two-phase currents flowing in a

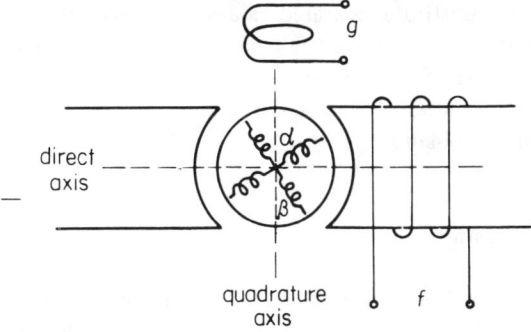

Figure 7.1 Slipring machine model

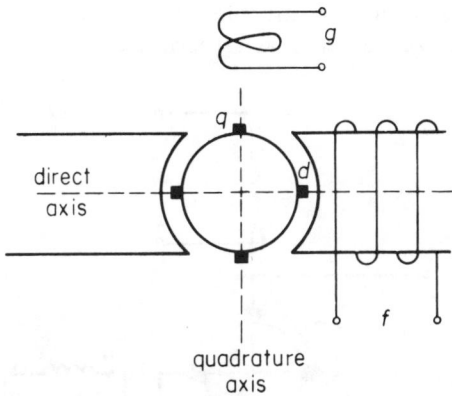

Figure 7.2 Commutator machine model

two-phase winding which will produce exactly the same magnetic field in the machine airgap.

An important restriction of the generalised theory is that only the stator can have salient poles, and the windings on the rotor must be balanced [3]. If necessary the physical form of a machine must be inverted so that it can be represented in this way. The conventional DC, synchronous and induction machines meet these requirements, as do many forms of DC and AC machine that are now obsolete. Unfortunately stepper motors are excluded because they have saliency on both the rotor and the stator. A further restriction is that magnetic non-linearity is neglected in the formulation of the equations. But this does not prevent saturation from being included; Jones [4] has shown that it is generally permissible to derive the performance equations of a machine on the assumption of linearity, and to substitute saturated values of the parameters at the end of the analysis. Sometimes this method fails, and non-linearity must be taken into account from the start [5]; an example is the calculation of the short-circuit current of a saturated alternator; but the simpler method is applicable to the majority of machine problems.

7.2 DC machine equations

Figure 7.3 is a symbolic representation of the commutator machine model shown in figure 7.2. It is a two-pole DC machine, with the normal field winding f wound on salient poles; the magnetic axis of these poles is known as the direct axis. The normal pair of brushes q lies on an axis known as the quadrature axis, at right angles to the direct axis. In addition, the model has a second field winding g on the quadrature axis, and a second pair of brushes d on the direct axis. This model represents the structure of the obsolete cross-field machines [6], with the ordinary DC machine as a special case.

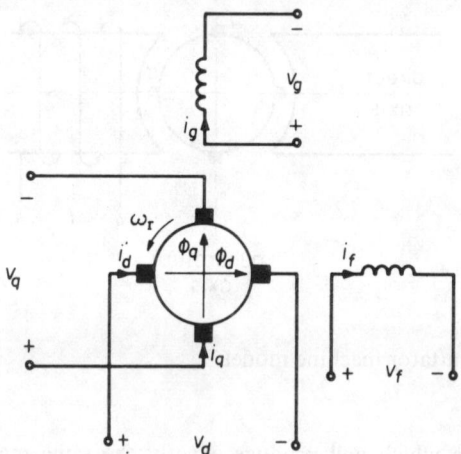

Figure 7.3 Symbolic representation of the commutator model

The currents in the brushes and the field windings will set up fluxes Φ_d and Φ_q on the direct and quadrature axes. As in the normal DC machine, rotation of the armature will generate a voltage between the q brushes on account of Φ_d, and a voltage between the d brushes on account of Φ_q. If the fluxes Φ_d and Φ_q vary with time, there will be induced voltages in the d and q windings respectively, owing to the rate of change of flux linkage; there will be no interaction between the axes, since they are at right angles.

Voltage equation for a field winding

For the main field winding on the direct axis we have

$$v_f = R_f i_f + \frac{d\psi_f}{dt}$$

$$= R_f i_f + \frac{d}{dt}(L_f i_f + M_{df} i_d)$$

$$= (R_f + L_f p)i_f + M_{df} p i_d \qquad (7.1)$$

where ψ_f is the flux linking the d-axis field winding (f), L_f is the self-inductance of the winding, and M_{df} is its mutual inductance with the d-axis armature circuit. The Heaviside notation of p for the d/dt operator is used rather than D, to avoid confusion with d which denotes the direct axis.

Voltage equation for an armature winding

For the brushes on the direct axis, we have

$$v_d = R_d i_d + \frac{d\psi_d}{dt} + K_a \Phi_q \omega_r \qquad (7.2)$$

where ω_r is the angular velocity of the rotor, ψ_d is the flux linking the d-axis armature circuit and K_a is the fundamental armature constant. The flux Φ_q will depend on i_q and i_g, and with normal machine windings Jones [4] has shown that

$$K_a \Phi_q = L_q i_q + M_{qg} i_g \qquad (7.3)$$

where L_q is the self-inductance of the q-axis armature circuit, and M_{qg} is the mutual inductance between this circuit and the q-axis field winding (g). Equation (7.2) therefore becomes

$$v_d = (R_d + L_d p)i_d + M_{df} p i_f + L_q \omega_r i_q + M_{qg} \omega_r i_g \qquad (7.4)$$

Matrix voltage equation

Expressions similar to eqns (7.1) and (7.4) may be written for the voltages v_g and v_q, and the set of four equations may then be put in matrix form

$$\begin{bmatrix} v_d \\ v_q \\ v_f \\ v_g \end{bmatrix} = \begin{bmatrix} R_d + L_d p & L_q \omega_r & M_{df} p & M_{qg} \omega_r \\ -L_d \omega_r & R_q + L_q p & -M_{df} \omega_r & M_{qg} p \\ M_{df} p & 0 & R_f + L_f p & 0 \\ 0 & M_{qg} p & 0 & R_g + L_g p \end{bmatrix} \begin{bmatrix} i_d \\ i_q \\ i_f \\ i_g \end{bmatrix} \qquad (7.5)$$

On account of the symmetry of the armature, $R_d = R_q = R_a$. But the magnetic circuits of the two axes are not in general identical, so that $L_d \neq L_q$. We may write the machine equations (7.5) in the form

$$v = Zi$$

and the square matrix Z is known as the *impedance matrix* of the machine.

7.3 AC machine equations

Consider the slipring model shown in figures 7.1 and 7.4. This has balanced two-phase windings α and β on the rotor, a main field winding f on the stator direct axis, and a second field winding g on the stator quadrature axis. With the winding omitted this represents a simple synchronous machine; with no saliency (a uniform airgap) and identical f and g windings, the model represents the two-phase induction machine. We may write the voltage equation for the α phase winding as

$$v_\alpha = R_\alpha i_\alpha + \frac{d\psi_\alpha}{dt}$$

$$= R_\alpha i_\alpha + p(L_\alpha i_\alpha + M_{\alpha\beta} i_\beta + M_{\alpha f} i_f + M_{\alpha g} i_g) \tag{7.6}$$

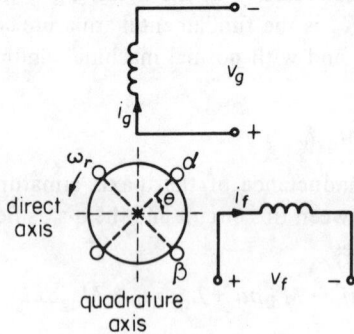

Figure 7.4   Symbolic representation of the slipring model

Variable inductance coefficients

In general, L_α, $M_{\alpha\beta}$, $M_{\alpha f}$ and $M_{\alpha g}$ are periodic functions of θ; to a first approximation we may take the first terms in the Fourier expansions of these functions. Thus L_α, the self-inductance of the winding, will vary from a maximum value L_d when the winding is aligned with the direct axis ($\theta = 0$ or π), to a minimum value L_q when the winding is aligned with the quadrature axis. The simplest sine or cosine function to meet this requirement is

$$L_\alpha = \tfrac{1}{2}(L_d + L_q) + \tfrac{1}{2}(L_d - L_q)\cos 2\theta \tag{7.7}$$

and this is equivalent to the assumption of a sinusoidally distributed winding – a good approximation which we have already used. The mutual inductance $M_{\alpha\beta}$ likewise is periodic in 2θ, and it may be shown that

$$M_{\alpha\beta} = \tfrac{1}{2}(L_d - L_q)\sin 2\theta \tag{7.8}$$

The mutual inductances $M_{\alpha f}$ and $M_{\alpha g}$ are clearly periodic in θ, and we may put

$$M_{\alpha f} = M_{df}\cos\theta \tag{7.9}$$

$$M_{\alpha g} = M_{qg}\sin\theta \tag{7.10}$$

In these equations, M_{df} is the maximum value of $M_{\alpha f}$ when the α phase is aligned with the d axis, and $M_{\alpha g}$ is the maximum value of $M_{\alpha g}$ when the α phase is aligned with the q axis. Equations (7.7) and (7.8) may be rewritten as

$$L_\alpha = L_d \cos^2\theta + L_q \sin^2\theta \tag{7.11}$$

$$M_{\alpha\beta} = (L_d - L_q)\sin\theta \cos\theta \tag{7.12}$$

and the voltage equation (7.6) becomes

$$v_\alpha = R_\alpha i_\alpha + p(L_d \cos^2\theta + L_q \sin^2\theta)i_\alpha + (L_d - L_q)p\sin\theta\cos\theta\, i_\beta + \\ + M_{df}p\cos\theta\, i_f + M_{qg}p\sin\theta\, i_g \tag{7.13}$$

Matrix voltage equation

The voltage equations for the other three windings can be derived in the same way, giving the following matrix equation

$$\begin{bmatrix} v_\alpha \\ v_\beta \\ v_f \\ v_g \end{bmatrix} = \begin{bmatrix} R_\alpha + p(L_d\cos^2\theta + L_q\sin^2\theta) & (L_d - L_q)p\sin\theta\cos\theta & M_{df}p\cos\theta & M_{qg}p\sin\theta \\ (L_d - L_q)p\sin\theta\cos\theta & R_\beta + p(L_d\sin^2\theta + L_q\cos^2\theta) & M_{df}p\sin\theta & M_{qg}p\cos\theta \\ M_{df}p\cos\theta & M_{df}p\sin\theta & R_f + L_f p & \\ M_{qg}p\sin\theta & M_{qg}p\cos\theta & & R_g + L_g p \end{bmatrix} \begin{bmatrix} i_\alpha \\ i_\beta \\ i_f \\ i_g \end{bmatrix}$$

$$(7.14)$$

In these equations, $R_\alpha = R_\beta = R_a$ since the armature winding is balanced.

Commutator transformation

To simplify the equations we introduce a change of variables which will remove the circular functions of θ from the impedance matrix. There are many possible transformations which will achieve this result; we select a transformation such that voltage and current transform in the same way, and the total instantaneous power (Σvi) is unchanged by the transformation. It may be shown [3] that a transformation to meet these requirements is given by

$$\begin{bmatrix} v_d \\ v_q \end{bmatrix} = \begin{bmatrix} \cos\theta & \sin\theta \\ \sin\theta & -\cos\theta \end{bmatrix} \begin{bmatrix} v_\alpha \\ v_\beta \end{bmatrix} \tag{7.15}$$

$$\begin{bmatrix} i_d \\ i_q \end{bmatrix} = \begin{bmatrix} \cos\theta & \sin\theta \\ \sin\theta & -\cos\theta \end{bmatrix} \begin{bmatrix} i_\alpha \\ i_\beta \end{bmatrix} \tag{7.16}$$

The square matrix in eqns (7.15) and (7.16) is termed the *commutator transformation matrix*, for a reason that will be explained shortly. If we denote this matrix by C, it has the useful property that $C = C^T = C^{-1}$; thus the same transformation gives the α,β quantities in terms of the d,q quantities. On making the substitutions indicated by eqns (7.15) and (7.16), eqn (7.14) becomes

$$\begin{bmatrix} v_d \\ v_q \\ v_f \\ v_g \end{bmatrix} \begin{bmatrix} R_a + L_d p & L_q \omega_r & M_{df} p & M_{qg} \omega_r \\ -L_d \omega_r & R_a + L_q p & -M_{df} \omega_r & M_{qg} p \\ M_{df} p & 0 & R_f + L_f p & 0 \\ 0 & M_{qg} p & 0 & R_g + L_g p \end{bmatrix} \begin{bmatrix} i_d \\ i_q \\ i_f \\ i_g \end{bmatrix} \tag{7.17}$$

This is identical to eqn (7.5) for the DC machine. Thus the transformation to d–q variables is equivalent to replacing the two-phase armature winding by a commutator winding with brushes on the d and q axes. The currents i_d and i_q are the currents which would have to flow in the brushes to give the same machine performance. An AC machine represented in d–q variables is shown symbolically in figure 7.5.

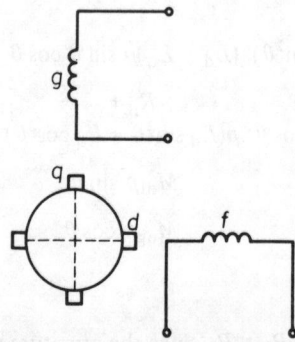

Figure 7.5 Two-axis machine model

7.4 General equations

General voltage equation

We have seen that the equations for the DC machine and for the AC machine in terms of d-q variables have the same form, and may be written as

$$v = Zi \qquad (7.18)$$

Three different kinds of quantity occur in the elements of the impedance matrix Z: constants, which denote the winding resistances; self or mutual inductance terms, of the form Lp; and generated voltage terms, of the form $L\omega_r$. The impedance matrix may thus be expressed in terms of three matrices:

$$Z = R + Lp + G\omega_r \qquad (7.19)$$

where R, L and G all have constant elements, and are given by

$$R = \begin{bmatrix} R_a & & & \\ & R_a & & \\ & & R_f & \\ & & & R_g \end{bmatrix}$$

$$L = \begin{bmatrix} L_d & 0 & M_{df} & 0 \\ 0 & L_q & 0 & M_{qg} \\ M_{df} & 0 & L_f & 0 \\ 0 & M_{qg} & 0 & L_g \end{bmatrix}$$

$$G = \begin{bmatrix} 0 & L_q & 0 & M_{qg} \\ -L_d & 0 & -M_{df} & 0 \\ 0 & 0 & 0 & 0 \\ 0 & 0 & 0 & 0 \end{bmatrix}$$

The resolution of Z into three parts reveals a simple structure, and Hancock [3] gives the following rules for writing down *by inspection* the impedance matrix for a machine with any number of windings

(1) Write in the principal diagonal the terms representing the resistance of the windings.
(2) Also along the principal diagonal write the Lp terms corresponding to the self-inductances of the windings.
(3) Write in the appropriate places the mutual inductance terms Mp wherever

the windings have a common axis. Each term will appear twice since mutual terms are always symmetric.

(4) Wherever Lp or Mp occurs in a d or q row, write $L\omega_r$ or $M\omega_r$ in the same column of the other q or d row, prefixing those in the q row with a negative sign.

Power in terms of matrices

For the DC machine in section 7.2, the total instantaneous power input to the windings is

$$P = i_d v_d + i_q v_q + i_f v_f + i_g v_g \qquad (7.20)$$

This may be expressed as a product of the voltage matrix with the transposed current matrix

$$P = \begin{bmatrix} i_d & i_q & i_f & i_g \end{bmatrix} \begin{bmatrix} v_d \\ v_q \\ v_f \\ v_g \end{bmatrix} \qquad (7.21)$$

$$= i^T v$$

For the AC machine of section 7.3, the instantaneous input power is likewise $P = i^T v$, where the elements of i and v are the actual winding quantities. The commutator transformation was chosen to satisfy the condition

$$i_\alpha v_\alpha + i_\beta v_\beta = i_d v_d + i_q v_q \qquad (7.22)$$

so the power is also $P = i^T v$ in terms of the d–q variables.

Torque equation

In terms of the components of Z, eqn (7.18) becomes

$$v = Ri + Lpi + G\omega_r i \qquad (7.23)$$

Pre-multiply this equation by i^T

$$i^T v = i^T R i + i^T L p i + i^T G \omega_r i \qquad (7.24)$$

In eqn (7.24), the terms have the following significance

$i^T v$ is the total instantaneous input power.

$i^T R i$ is of the form $\Sigma R i^2$ since R is diagonal; it therefore represents the total ohmic loss in the machine.

$i^{\mathrm{T}}Lpi$ is the sum of terms of the form $iLpi = p(\tfrac{1}{2}Li^2)$ and $i_1 M p i_2 + i_2 M p i_1 = p(M i_1 i_2)$; it therefore represents the rate of change of stored magnetic energy.

$i^{\mathrm{T}}G\omega_r i$ is the difference between the input power and the other two terms; it represents the mechanical output power.

If T is the electromagnetic torque, then the mechanical output power is $T\omega_r$; thus

$$T\omega_r = i^{\mathrm{T}} G \omega_r i$$

giving

$$T = i^{\mathrm{T}} G i \tag{7.25}$$

This is one form of the general torque equation for electrical machines. As a particular case, take the simple DC machine. The impedance matrix is obtained by deleting the g and d rows and columns from eqn (7.5)

$$Z = \begin{bmatrix} R_a + L_q p & -M_{df}\omega_r \\ 0 & R_f + L_f p \end{bmatrix} \tag{7.26}$$

Thus

$$G = \begin{bmatrix} 0 & -M_{df} \\ 0 & 0 \end{bmatrix} \tag{7.27}$$

giving

$$T = i^{\mathrm{T}} G i = -M_{df} i_f i_q \tag{7.28}$$

which is equivalent to the equation

$$T = K i_f i_a \tag{2.14}$$

developed earlier for the simple DC machine.

7.5 Applications

Simple DC machines seldom require the generalised approach, and the cross-field machines which do require it are obsolete. In this section we show how the generalised theory can be applied to three AC machine problems: the steady-state performance of synchronous and induction machines, and the dynamic braking of an induction motor.

Salient-pole synchronous machine

Consider a machine with only one field winding, placed on the d axis, and no damper windings (figure 7.6). The matrix voltage equation is obtained from

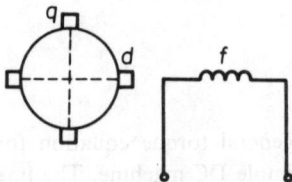

Figure 7.6 Synchronous machine without damper windings

eqn (7.17) by omitting the g row and column; this eliminates the mutual inductance M_{qg} and the remaining mutual inductance M_{df} may be written as M

$$\begin{bmatrix} v_d \\ v_q \\ v_f \end{bmatrix} = \begin{bmatrix} R_a + L_d p & L_q \omega_r & Mp \\ -L_d \omega_r & R_a + L_q p & -M\omega_r \\ Mp & 0 & R_f + L_f p \end{bmatrix} \begin{bmatrix} i_d \\ i_q \\ i_f \end{bmatrix} \quad (7.29)$$

For balanced steady-state operation, the phase voltages and currents will be sinusoidal quantities. If the physical machine has three phases then from appendix B we have the following equivalent two-phase quantities

$$v_\alpha = \sqrt{3} V \sin \omega t \quad (7.30)$$

$$v_\beta = \sqrt{3} V \sin(\omega t - \pi/2) = -\sqrt{3} V \cos \omega t \quad (7.31)$$

$$i_\alpha = \sqrt{3} I \sin(\omega t - \phi) \quad (7.32)$$

$$i_\beta = \sqrt{3} I \sin(\omega t - \phi - \pi/2) = -\sqrt{3} I \cos(\omega t - \phi) \quad (7.33)$$

where V and I are the RMS values of the three-phase armature voltage and current. The armature rotates with a steady angular velocity $\omega_r = \omega$; since $\omega_r = d\theta/dt$, we have

$$\theta = \omega t - \delta \quad (7.34)$$

where δ is the angle between the axis of the α phase and the d axis at time $t = 0$.

Two-axis voltage equations

The commutator transformation (eqn 7.15) gives

$$\begin{bmatrix} v_d \\ v_q \end{bmatrix} = \begin{bmatrix} \cos(\omega t - \delta) & \sin(\omega t - \delta) \\ \sin(\omega t - \delta) & -\cos(\omega t - \delta) \end{bmatrix} \begin{bmatrix} \sqrt{3}V \sin \omega t \\ -\sqrt{3}V \cos \omega t \end{bmatrix} = \begin{bmatrix} \sqrt{3}V \sin \delta \\ \sqrt{3}V \cos \delta \end{bmatrix}$$
(7.35)

Similarly

$$i_d = \sqrt{3}I \sin(\delta - \phi) \tag{7.36}$$

$$i_q = \sqrt{3}I \cos(\delta - \phi) \tag{7.37}$$

Thus v_d, v_q, i_d and i_q are all steady (DC) quantities. For steady-state operation v_f and i_f will both be steady DC quantities; all the time derivatives in eqn (7.29) will vanish, and the equation may be written in the form

$$\begin{bmatrix} \sqrt{3}V \sin \delta \\ \sqrt{3}V \cos \delta \\ V_f \end{bmatrix} = \begin{bmatrix} R_a & X_q & 0 \\ -X_d & R_a & -X_m \\ 0 & 0 & R_f \end{bmatrix} \begin{bmatrix} \sqrt{3}I \sin(\delta - \phi) \\ \sqrt{3}I \cos(\delta - \phi) \\ I_f \end{bmatrix}$$
(7.38)

where $X_d = \omega L_d$, $X_q = \omega L_q$ and $X_m = \omega M$. Thus

$$V \sin \delta = R_a I \sin(\delta - \phi) + X_q I \cos(\delta - \phi) \tag{7.39}$$

$$V \cos \delta = -X_d I \sin(\delta - \phi) + R_a I \cos(\delta - \phi) - X_m I_f/\sqrt{3} \tag{7.40}$$

With the armature on open circuit, $I = 0$; and from eqn (7.39), $\delta = 0$. If we let $V = E$ under these conditions, then eqn (7.40) gives

$$E = [V]_{I=0} = -X_m I_f/\sqrt{3} \tag{7.41}$$

Equations (7.39) and (7.40) now become

$$V \sin \delta = X_q I \cos(\delta - \phi) + R_a I \sin(\delta - \phi) \tag{7.42}$$

$$V \cos \delta = E - X_d I \sin(\delta - \phi) + R_a I \cos(\delta - \phi) \tag{7.43}$$

These are the basic equations for the steady-state operation of a synchronous machine. If we neglect armature resistance by putting $R_a = 0$, and remove saliency by letting $X_d = X_q = X_s$, then eqns (7.42) and (7.43) reduce to

$$V \sin \delta = X_s I \cos(\delta - \phi) \tag{7.44}$$

$$V \cos \delta = E - X_s I \sin(\delta - \phi) \tag{7.45}$$

These equations may be obtained from the phasor diagram for the synchronous machine derived earlier (figure 5.8).

Torque equation

The torque is given by eqn (7.25); from eqn (7.29) the matrix G is

$$G = \begin{bmatrix} 0 & L_q & 0 \\ -L_d & 0 & -M \\ 0 & 0 & 0 \end{bmatrix} \quad (7.46)$$

and the torque is therefore

$$T = i^T G i = L_q i_q i_d - L_d i_d i_q - M i_f i_q$$

$$= \frac{1}{\omega}\{i_d i_q (X_q - X_d) - i_f i_q X_m\} \quad (7.47)$$

If R_a can be neglected, eqns (7.42) and (7.43) become

$$V \sin \delta = X_q I \cos(\delta - \phi) \quad (7.48)$$

$$V \cos \delta = E - X_d I \sin(\delta - \phi) \quad (7.49)$$

From eqns (7.36) and (7.37), these equations may be written as

$$\sqrt{3} V \sin \delta = X_q i_q \quad (7.50)$$

$$\sqrt{3} V \cos \delta = \sqrt{3} E - X_d i_d \quad (7.51)$$

and

$$\sqrt{3} E = -X_m i_f \quad [7.41]$$

Substitution for i_d, i_q and i_f from these equations into eqn (7.47) gives the torque

$$T = \frac{3}{\omega}\left[\frac{VE}{X_d}\sin\delta + \tfrac{1}{2}V^2\left\{\frac{1}{X_q} - \frac{1}{X_d}\right\}\sin 2\delta\right] \quad (7.52)$$

If $X_d = X_q$, this equation reduces to the torque expression derived earlier for the cylindrical rotor synchronous machine (eqn 5.8), with $m = 3$ and $p = 1$.

Induction machine: steady-state operation

A three-phase induction machine may be represented by a two-phase model as shown in figure 7.4. If balanced three-phase voltages are applied to the stator, then the three-phase to two-phase transformation gives

$$v_f = \sqrt{3} V \cos \omega t$$

$$v_g = \sqrt{3} V \cos(\omega t - \pi/2) = \sqrt{3} V \sin \omega t$$

The rotor windings are short-circuited; consequently $v_\alpha = v_\beta = 0$, and the commutator transformation gives $v_d = v_q = 0$. Now consider the two-axis model of figure 7.5. Since the two stator phases f and g are balanced, and there is no saliency, the machine is completely symmetrical about the two axes; the general equations therefore take the following simple form

$$\begin{bmatrix} 0 \\ 0 \\ v_f \\ v_g \end{bmatrix} = \begin{bmatrix} R_2 + L_2 p & L_2 \omega_r & Mp & M\omega_r \\ -L_2 \omega_r & R_2 + L_2 p & -M\omega_r & Mp \\ Mp & 0 & R_1 + L_1 p & 0 \\ 0 & Mp & 0 & R_1 + L_1 p \end{bmatrix} \begin{bmatrix} i_d \\ i_q \\ i_f \\ i_g \end{bmatrix} \quad (7.53)$$

If the rotor speed ω_r is constant, this is a set of linear differential equations with constant coefficients. Since v_f and v_g are sinusoidal, the currents i_d, i_q, i_f, i_g must also be sinusoidal quantities with the same angular frequency ω. We may therefore transform the equations from the time domain to the frequency domain, in the normal manner of AC circuit theory; time-varying quantities are replaced by complex (phasor) quantities, and the d/dt operator p is replaced by $j\omega$

$$\begin{bmatrix} 0 \\ 0 \\ V_1 \\ -jV_1 \end{bmatrix} = \begin{bmatrix} R_2 + jX_2 & vX_2 & jX_m & vX_m \\ -vX_2 & R_2 + jX_2 & -vX_m & jX_m \\ jX_m & 0 & R_1 + jX_1 & 0 \\ 0 & jX_m & 0 & R_1 + jX_1 \end{bmatrix} \begin{bmatrix} I_d \\ I_q \\ I_f \\ I_g \end{bmatrix} \quad (7.54)$$

where $X_1 = \omega L_1$, $X_2 = \omega L_2$, $X_m = \omega M$, $v = \omega_r/\omega$, and $V_1 = \sqrt{(3/2)} V$.

Equation (7.54) can be solved for the unknown currents [3], but it is simpler to make use of the two-phase symmetry and assume that the currents are given by

$$I_d = I_2, \quad I_q = -jI_2, \quad I_f = I_1, \quad I_g = -jI_1$$

Expansion of eqn (7.54) then yields two independent equations

$$0 = (R_2 + jsX_2)I_2 + jsX_m I_1 \quad (7.55)$$
$$V_1 = jX_m I_2 + (R_1 + jX_1)I_1 \quad (7.56)$$

where $s = 1 - v = (\omega - \omega_r)/\omega$ is the fractional slip. Equation (7.55) shows that $I_2 = 0$ when $s = 0$; we may divide through by s, giving

$$0 = (R_2/s + jX_2)I_2 + jX_m I_1 \quad (7.57)$$

As with the transformer, leakage reactances may be defined as follows

$$x_1 = X_1 - X_m, \quad x_2 = X_2 - X_m$$

Equations (7.56) and (7.57) then become

$$V_1 = jX_m(I_1 + I_2) + (R_1 + jx_1)I_1 \qquad (7.58)$$

$$0 = (R_2/s + jx_2)I_2 + jX_m(I_1 + I_2) \qquad (7.59)$$

These are the equations of the induction-machine equivalent circuit shown in figure 6.12, if the core loss resistance R_c is omitted. This is an important result; not only does it give an alternative derivation of the equivalent circuit, but it also provides a simple method of finding the parameters of the impedance matrix. Standard tests on the induction machine [7] will yield values for R_1, R_2, X_m and $x_1 + x_2$. It may be shown that the ratio x_1/x_2 will alter the value of I_2, but that the calculated performance is independent of the ratio. We may therefore set $x_1 = x_2$; this defines X_1 and X_2, so all the parameters may be determined.

Induction motor: dynamic braking

An induction motor can be stopped rapidly by disconnecting the AC supply from the stator and substituting a DC source; this is termed *dynamic braking* (see problem 6.3). One method of connecting the windings to a voltage source is shown in figure 7.7; the equivalent two-phase stator voltages are then given by

$$v_f = \sqrt{(2/3)}V, \quad v_g = 0$$

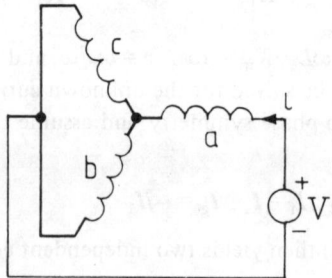

Figure 7.7 Stator connection for dynamic braking of an induction motor

These values may be substituted in eqn (7.53) to give the electrical differential equations of the machine. For the dynamic braking condition, ω_r is not constant; consequently the equations are non-linear. If there is no mechanical load on the motor, all the developed torque will be absorbed in the acceleration of the rotor. The torque is given by eqn (7.25), and expansion of this equation for the induction motor gives

$$T = M(i_d i_g - i_q i_f) \qquad (7.60)$$

Thus the rotor equation of motion is

$$J \frac{d\omega_r}{dt} = T = M(i_d i_g - i_q i_f) \tag{7.61}$$

where J is the rotor moment of inertia. Solution of these non-linear equations is easily accomplished numerically if they can be rearranged with all the time derivatives on the left-hand side; standard numerical algorithms are available for equations in this form. From eqn (7.23) the electrical equations have the form

$$v = (R + Lp + G\omega_r)i$$

Thus

$$pi = L^{-1}\{v - (R + G\omega_r)i\} \tag{7.62}$$

Equation (7.61) may be written in the form

$$p\omega_r = J^{-1} M(i_d i_g - i_q i_f) \tag{7.63}$$

and eqns (7.62) and (7.63) are now in the required form for solution. For the induction motor, the inverse of the inductance matrix takes the simple form

$$L^{-1} = \begin{bmatrix} A & 0 & C & 0 \\ 0 & A & 0 & C \\ C & 0 & B & 0 \\ 0 & C & 0 & B \end{bmatrix} \tag{7.64}$$

where

$$A = L_1/D, \quad B = L_2/D, \quad C = -M/D, \quad D = L_1 L_2 - M^2$$

Appendix C gives a computer program for the solution of eqns (7.62) and (7.63), using the trapezoidal integration algorithm. Figure 7.8 is a graph of the computed rotor speed against time for a 3 kW cage induction motor, connected as shown in figure 7.7 to a DC supply of 20 V. The rated phase voltage for the motor is 230 V, the rotor moment of inertia is 0.011 kg m^2, and the parameters determined by test are as follows

$$R_1 = 1.80 \, \Omega, \quad R_2 = 1.81 \, \Omega, \quad L_1 = L_2 = 0.249 \, \text{H}, \quad M = 0.240 \, \text{H}$$

An interesting feature of dynamic braking, which can be seen in figure 7.8, is that the rotor actually reverses before it finally comes to rest. This happens because induced currents continue to flow in the inductance of the rotor circuit when the speed first reaches zero, and the resulting torque reverses the rotor.

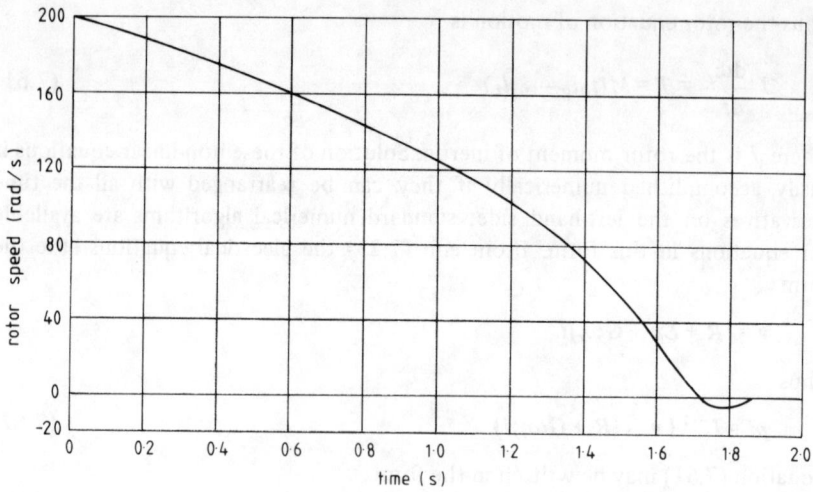

Figure 7.8 Computed rotor speed for dynamic braking of an induction motor

7.6 Conclusion

The generalised theory outlined in this chapter is particularly useful in the more complex machine problems, such as the transient and unbalanced operation of AC machines. It is the basis for investigating the stability of machines [3, 8, 9], and for analysing systems such as variable-speed drives. Equally important is the unification of electrical machine theory; the same equations and methods apply to all machines which satisfy the basic assumptions of the theory, and this includes the conventional DC and AC machines. Of course there are limits to the scope of the generalised theory; we have already noted that doubly-salient machines such as stepper motors cannot be analysed in this way; also excluded are machines such as linear induction motors and thick-cylinder induction motors, which require field theory rather than circuit theory for their analysis.

Two features of the generalised theory should be noted. One is the generality of the basic equations; they apply to all conditions of operation, with the steady state as a particular case. The second feature is the systematic way in which the equations are formulated. The impedance matrix for a machine with any number of windings can be written down by inspection, with the known structure of the matrix forming an unequivocal check of its correctness. Interconnection of the windings (or of several machines) can also be handled with matrix algebra [3] in a routine process which is easily checked. It is thus possible to arrive at the final equations which specify the performance of a new machine or machine system with the certainty that these equations are correct. The solution of the equations is, of course, another matter. Analytical solution may be difficult when the equations are linear, and impossible when they are non-linear; but numerical solution is relatively simple (see the example of dynamic braking in section 7.5). When they can be obtained, analytical solutions are valuable because of the insight which a formula can give. Numerical methods yield solutions to particular problems with ease – they can be implemented on microcomputers – but lack the generality of an analytical formula. Nevertheless, numerical methods enable complex practical problems to be solved, and this makes the generalised theory a powerful tool for the application engineer.

Problems

7.1. In the sinusoidal steady state, the time-domain matrix voltage equation (7.18) becomes

$$V = ZI = (R + j\omega L + G\omega_r)I$$

where the elements of V and I are complex (phasor) quantities. Show that the average torque is given by

$$T_{av} = \text{Re}(I^{*T}GI)$$

where I^* is the complex conjugate of I.

7.2. Use the result of problem 7.1 and the induction machine equations in section 7.5 to show that the torque developed by a three-phase induction machine is

$$T_{av} = 3(I_2')^2 R_2/s$$

where I_2' is the RMS current in the rotor branch of the equivalent circuit for one phase.

7.3. The damper windings of a synchronous machine can be represented by short-circuited windings k_d and k_q on the direct and quadrature axes of the stator in the two-axis model. Write down by inspection the matrix voltage equation for the machine. For steady-state operation only, show that the damper windings will carry no current and that the corresponding rows and columns can be deleted from the impedance matrix.

7.4. A synchronous motor with damper windings is started by induction action, with the field winding short-circuited. If the rotor revolves with a constant angular velocity ω_r, so that $\theta = \omega_r t - \delta$, show that the two-axis components of armature voltage are given by

$$v_d = \sqrt{3}V \sin(s\omega t + \delta)$$
$$v_q = \sqrt{3}V \cos(s\omega t + \delta)$$

where $s = (\omega - \omega_r)/\omega$. Deduce that all the two-axis currents will alternate with angular frequency $s\omega$, and show that

(a) the phase current will contain components with angular frequencies ω and $(1 - 2s)\omega$
(b) the torque will contain a constant component and an oscillatory component with angular frequency $2s\omega$.

Hint: it is not necessary to solve any equations; let the currents be $i_d = I_{dm} \sin(s\omega t - \phi_d)$, and so on; in part (b) make use of the fact that the torque can only contain the products of pairs of current, and expand a typical product.

References

1 G. Kron, *The Application of Tensors to the Analysis of Rotating Electrical Machinery* (Schenectady: General Electric Review, 1938).
2 G. Kron, *Tensors for Circuits*, 2nd ed. (New York: Dover, 1959).
3 N. N. Hancock, *Matrix Analysis of Electrical Machinery*, 2nd ed. (Oxford: Pergamon Press, 1974).
4 C. V. Jones, *The Unified Theory of Electrical Machines* (London:Butterworths, 1967).
5 I. G. Reddy, A. M. Ali and C. V. Jones, 'Computer-aided analysis of saturated systems', *Proc. IEE*, **118** (1971), pp. 1791-9.
6 M. G. Say and E. O. Taylor, *Direct-Current Machines* (London: Pitman, 1980).
7 P. L. Alger, *Induction Machines*, 2nd ed. (New York: Gordon and Breach, 1970).
8 G. J. Rogers, 'Linearised analysis of induction motor transients', *Proc. IEE*, **112** (1965), pp. 1917-26.
9 P. J. Lawrenson and S. R. Bowes, 'Stability of reluctance machines', *Proc. IEE*, **118** (1971), pp. 356-69.

Appendix A: Airgap Field Components and the Maxwell Stress

Figure A.1 shows an idealised machine structure. The stator and rotor surfaces are smooth; the permeability of the iron is assumed to be infinite; and the windings are represented by 'current sheets' of negligible thickness on the stator and rotor surfaces. Current flows in the axial direction, perpendicular to the plane of the paper. Let the stator linear current density be

$$K_1 = -K_{1m} \sin \theta \tag{A.1}$$

The radial and circumferential components of the magnetic field may be found by applying Ampère's circuital law to selected paths. The radial component H_{1r} may be evaluated from a path such as PQRS (figure A.1), as was done in section 4.2; from eqns (4.9) to (4.11), the result is

$$H_{1r} = \frac{r_1}{g} K_{1m} \cos \theta \tag{A.2}$$

where r_1 is the radius of the stator current sheet and g is the radial length of the airgap; it is assumed that H_{1r} is constant along a radial path such as PQ.

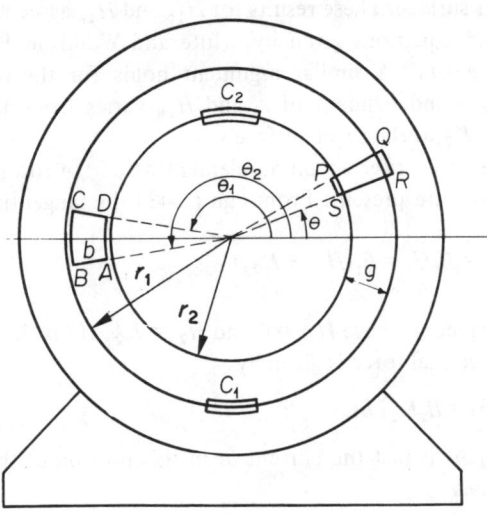

Figure A.1

Next we show that there must be a tangential component of magnetic field. Consider a path in the airgap which links no current, such as ABCD in figure A.1. We have

$$0 = \oint H_1 \cdot ds = \int_B^C H_{1s} \, ds + \int_D^A H_{1s} \, ds + \frac{b}{g} r_1 K_{1m} (\cos \theta_2 - \cos \theta_1) \quad (A.3)$$

where $AB = CD = b$. Equation (A.3) shows that there must be a tangential component H_{1s} along BC or AD. To evaluate this component, consider the boundary conditions at the stator and rotor surfaces. Take closed contours C_1 and C_2 enclosing lengths δs_1 and δs_2 of the respective surfaces; let the ends of the contours shrink to zero in such a way that one curved side is just in the iron, while the other curved side is just in the air. For the contour C_1, the current enclosed is $K_1 \, \delta s_1$, and Ampère's circuital law gives

$$K_1 \, \delta s_1 = \oint_{C_1} H_1 \cdot ds = H_{1s} \, \delta s_1 \quad (A.4)$$

Thus $H_{1s} = K_1$ on the airgap side of the stator surface. Since we are considering the field due to stator current alone, the contour C_2 encloses no current, and we have

$$0 = \oint_{C_2} H_1 \cdot ds = H_{1s} \, \delta s_2 \quad (A.5)$$

Thus $H_{1s} = 0$ on the airgap side of the rotor surface. If H_{1r} is independent of r, and the airgap length g is small in comparison with the radius r_1, it is readily shown that H_{1s} varies linearly with r from a value of 0 at the rotor surface to K_1 at the stator surface. These results for H_{1r} and H_{1s} agree with the exact solution of the field equations given by White and Woodson [1], subject to the condition that $g \ll r_1$. A similar argument holds for the rotor field H_2; the component H_{2r} is independent of r, and H_{2s} varies from a value of 0 at the stator surface to K_2 at the rotor surface.

Consider the force exerted on an element δs of the rotor when both stator and rotor currents are present. From eqn (1.43) the tangential Maxwell stress is

$$t_s = \frac{B_r B_s}{\mu_0} = B_r H_s = B_r (H_{1s} + H_{2s}) \quad (A.6)$$

At the rotor surface we have $H_{1s} = 0$ and $H_{2s} = K_2$. If l is the axial length of the element, the tangential force is given by

$$\delta F_s = t_s l \, \delta s = B_r K_2 l \, \delta s \quad (A.7)$$

The quantity $k_2 \, \delta s$ is just the current δi in this portion of the rotor surface, so eqn (A.7) becomes

$$\delta F_s = B_r l \, \delta i \quad (A.8)$$

Equation (A.8) is equivalent to eqn (4.25), showing that the Maxwell stress calculation is equivalent to evaluating the force on a current element in a magnetic field.

Reference

1 D. C. White and H. H. Woodson, *Electromechanical Energy Conversion* (New York: Wiley, 1959).

Appendix B: Three-phase to Two-phase Transformation

We consider the conditions under which it is possible to replace a three-phase machine winding with a two-phase winding, so that the performance of the machine is unchanged. The two windings are shown symbolically in figure B.1,

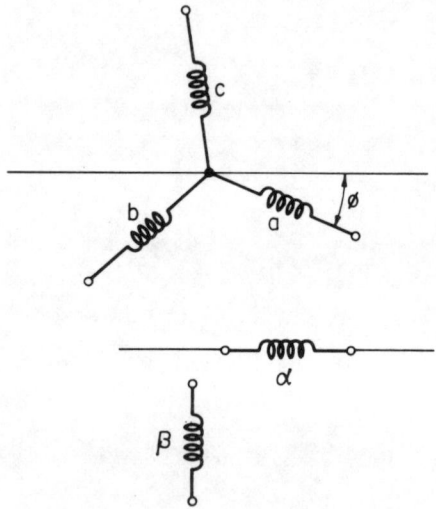

Figure B. 1

and they are assumed to be sinusoidally distributed. Let currents i_a, i_b, i_c flow in the three-phase winding; then the current density at an angle θ from the reference axis will be

$$K_3 = -Z_3\{i_a \sin(\theta + \phi) + i_b \sin(\theta + \phi + 2\pi/3) + i_c \sin(\theta + \phi + 4\pi/3)\} \quad \text{(B.1)}$$

If currents i_α, i_β flow in the two-phase winding, the corresponding current density will be

$$\begin{aligned}K_2 &= -Z_2\{i_\alpha \sin\theta + i_\beta \sin(\theta + \pi/2)\} \\ &= -Z_2(i_\alpha \sin\theta + i_\beta \cos\theta)\end{aligned} \quad \text{(B.2)}$$

APPENDIX B: THREE-PHASE TO TWO-PHASE TRANSFORMATION

The two windings will be equivalent if $K_3 = K_2$ for all values of θ. Equation (B.1) may be expanded to give

$$K_3 = -Z_3 i_a (\sin\theta \cos\phi - \cos\theta \sin\phi)$$
$$+ Z_3 i_b \{\sin\theta \cos(\phi + 2\pi/3) + \cos\theta \sin(\phi + 2\pi/3)\}$$
$$+ Z_3 i_c \{\sin\theta \cos(\phi + 4\pi/3) + \cos\theta \sin(\phi + 4\pi/3)\} \quad (B.3)$$

and the condition for equivalence is obtained by equating the coefficients of $\sin\theta$ and $\cos\theta$ in eqns (B.2) and (B.3). This gives

$$i_\alpha = \frac{Z_3}{Z_2}\{i_a \cos\phi + i_b \cos(\phi + 2\pi/3) + i_c \cos(\phi + 4\pi/3)\} \quad (B.4)$$

$$i_\beta = \frac{Z_3}{Z_2}\{i_a \sin\phi + i_b \sin(\phi + 2\pi/3) + i_c \sin(\phi + 4\pi/3)\}$$

These relations may be written in matrix form

$$\begin{bmatrix} i_\alpha \\ i_\beta \end{bmatrix} = k \begin{bmatrix} \cos\phi & \cos(\phi + 2\pi/3) & \cos(\phi + 4\pi/3) \\ \sin\phi & \sin(\phi + 2\pi/3) & \sin(\phi + 4\pi/3) \end{bmatrix} \begin{bmatrix} i_a \\ i_b \\ i_c \end{bmatrix} \quad (B.5)$$

where $k = Z_3/Z_2$. The corresponding relationship for voltage may be deduced if we stipulate that the total instantaneous input power is to be the same for both windings. We write eqn (B.5) in the form

$$i_2 = A i_3 \quad (B.6)$$

where i_2 is the column vector of two-phase currents, and i_3 is the vector of three-phase currents. Let v_2 and v_3 be the corresponding column vectors of voltages. Since the power is the same

$$i_3^T v_3 = i_2^T v_2 \quad (B.7)$$

Substituting for i_2 from eqn (B.6) gives

$$i_3^T v_3 = (A i_3)^T v_2 = i_3^T A^T v_2 \quad (B.8)$$

If eqn (B.8) is to hold for any arbitrary i_3, then

$$v_3 = A^T v_2 \quad (B.9)$$

Since the matrices A and A^T are singular, eqns (B.6) and (B.9) cannot be inverted as they stand. For a machine with no neutral connection, however, the three currents i_a, i_b, i_c are not all independent; we must have

$$i_a + i_b + i_c = 0 \quad (B.10)$$

showing that there are only two independent three-phase currents. For machines with balanced windings, it may be shown [1] that the phase voltages v_a, v_b, v_c are also related by the equation

$$v_a + v_b + v_c = 0 \tag{B.11}$$

when there is no neutral connection. Equation (B.11) is exactly true for machines with a uniform airgap, and a reasonable approximation for salient-pole machines. We may incorporate eqns (B.10) and (B.11) in eqns (B.6) and (B.9) by including additional elements in the matrices; thus

$$\begin{bmatrix} i_\alpha \\ i_\beta \\ 0 \end{bmatrix} = k \begin{bmatrix} \cos\phi & \cos(\phi + 2\pi/3) & \cos(\phi + 4\pi/3) \\ \sin\phi & \sin(\phi + 2\pi/3) & \sin(\phi + 4\pi/3) \\ m & m & m \end{bmatrix} \begin{bmatrix} i_a \\ i_b \\ i_c \end{bmatrix} \tag{B.12}$$

$$\begin{bmatrix} v_a \\ v_b \\ v_c \end{bmatrix} = k \begin{bmatrix} \cos\phi & \sin\phi & m \\ \cos(\phi + 2\pi/3) & \sin(\phi + 2\pi/3) & m \\ \cos(\phi + 4\pi/3) & \sin(\phi + 4\pi/3) & m \end{bmatrix} \begin{bmatrix} v_\alpha \\ v_\beta \\ 0 \end{bmatrix} \tag{B.13}$$

In these equations m is an arbitrary constant, and the factor $k = Z_3/Z_2$ can be assigned any value because Z_3 and Z_2 are arbitrary. Since the square transformation matrix is non-singular it can be inverted; if we write eqns (B.12) and (B.13) in the form

$$i'_2 = A' i_3 \tag{B.14}$$

$$v_3 = A'^T v'_2 \tag{B.15}$$

then the required inverse relationships are

$$i_3 = (A')^{-1} i'_2 \tag{B.16}$$

$$v'_2 = (A'^T)^{-1} v_3 \tag{B.17}$$

It would be very convenient if voltage and current transformed in the same way, that is if

$$A'^T = (A')^{-1} \tag{B.18}$$

and

$$A' = (A'^T)^{-1} \tag{B.19}$$

We thus require the matrix A' to be orthogonal, and this will be the case if $k = \sqrt{(2/3)}$ and $m = 1/\sqrt{2}$. Finally, we may omit the additional term from the matrices and for convenience let $\phi = 0$. Then

$$\begin{bmatrix} i_\alpha \\ i_\beta \end{bmatrix} = \sqrt{(2/3)} \begin{bmatrix} 1 & -1/2 & -1/2 \\ 0 & \sqrt{3}/2 & -\sqrt{3}/2 \end{bmatrix} \begin{bmatrix} i_a \\ i_b \\ i_c \end{bmatrix} \tag{B.20}$$

APPENDIX B: THREE-PHASE TO TWO-PHASE TRANSFORMATION

$$\begin{bmatrix} i_a \\ i_b \\ i_c \end{bmatrix} = \sqrt{(2/3)} \begin{bmatrix} 1 & 0 \\ -1/2 & \sqrt{3}/2 \\ -1/2 & -\sqrt{3}/2 \end{bmatrix} \begin{bmatrix} i_\alpha \\ i_\beta \end{bmatrix} \quad (B.21)$$

with similar equations for the phase voltages.

We can thus convert from three to two phases and vice versa for all conditions of operation – balanced or unbalanced, steady-state or transient – proved only that there is a three-wire connection to the three-phase winding. A three-phase machine may therefore be analysed in terms of the simpler two-phase model, and the results transformed back to three phases by means of eqn (B.21). A particular advantage of choosing $\phi = 0$ is that i_a is independent of i_β, and is simply equal to $\sqrt{(2/3)}i_\alpha$.

When the voltages or currents are balanced sinusoidal quantities, the transformations take a particularly simple form. For example, consider the three-phase voltages

$$v_a = \sqrt{2}V \cos \omega t$$
$$v_b = \sqrt{2}V \cos(\omega t - 2\pi/3) \quad (B.22)$$
$$v_c = \sqrt{2}V \cos(\omega t - 4\pi/3)$$

Application of the transformation in eqn (B.21) gives

$$v_\alpha = \sqrt{3}V \cos \omega t$$
$$v_\beta = \sqrt{3}V \cos(\omega t - \pi/2) \quad (B.23)$$
$$= \sqrt{3}V \sin \omega t$$

Reference

1 C. V. Jones, *The Unified Theory of Electrical Machines* (London: Butterworths, 1967).

Appendix C: Dynamic Braking of an Induction Motor

The computer program below carries out the numerical solution of the non-linear induction motor equations for dynamic braking (see section 7.5). It uses a subset of standard Pascal which should be available on most computers: only real and integer variables and real arrays are required, together with the standard Pascal control structures. A simple integration algorithm — the trapezium rule — is used for illustration in the program; in practice a more sophisticated algorithm from a program library would be used. The sample output was obtained with a Nascom 2 microcomputer (Z80 processor), using the Nascom Pascal compiler with a floating-point precision of approximately 11 decimal digits.

Only three features of the program require comment: the term *communication interval*, represented by the real variable 'comint'; the use of arrays for storing variables and derivatives; and the first part of the derivative procedure 'deriv'. The communication interval is the time between successive printing of the solution values; during this interval several integration steps may be taken in order to keep the truncation error small. Arrays 'x' and 'px' have been used for storing the five dependent variables and their derivatives so that a general-purpose integration procedure could be incorporated in the program. Data would need to be organised in this way if an integration procedure from an external library were to be used. In the derivative procedure, values in the dependent variable array are assigned to local variables, partly for efficiency but mainly to make the procedure easier to follow. The program is a straightforward implementation of the equations in section 7.5, and it is easily converted into Basic or Fortran.

Pascal program

```
program dynamic_braking;
{Numerical solution of the two-axis equations for
 dynamic braking of an induction motor.
 J D Edwards  09/04/85}

const
    nvar=5;pi=3.1415926535;
```

APPENDIX C: DYNAMIC BRAKING OF AN INDUCTION MOTOR

```
var
   R1,R2,L1,L2,M,J,V,w_ri,v_f,MJ,A,B,C,D,t,dt,comint,tmax:  real;
   x,px,w,pw: array[1 .. nvar] of real;
   i,steps: integer;

procedure get_parameters;
{input of machine and integration parameters}
begin
   writeln('Machine parameters:');writeeln;
   write('Primary resistance        R1 = '); readln(R1);
   write('Secondary resistance      R2 = '); readln(R2);
   write('Primary inductance        L1 = '); readln(L1);
   write('Secondary inductance      L2 = '); readln(L2);
   write('Mutual inductance         M = '); readln(M);
   write('Rotor moment of inertia   J = '); readln(J);
   write('Applied DC voltage        V = '); readln(V);
   write('Initial rotor speed       Wr = '); readln(w_ri);
   writeln;
   writeln('Integration parameters:'); writeln;
   write('Integration step length   = '); readln(dt);
   write('Communication interval    = '); readln(comint);
   write('Total integration time    = '); readln(tmax);
   steps :=round(comint/dt);
   MJ := M/J;
   D := L1*L2 - sqr (M) ;
   A := L1/D;
   B := L2/D;
   C := -M/D;
   v_f := sqrt(2/3)*V;
end; {of get_parameters}

procedure write_parameters;
{writes out machine and integration parameters}
begin
   writeln('Machine parameters: ') ;writeln;
   writeln('Primary resistance        R1 =' ,R1:5:2,' ohms') ;
   writeln('Secondary resistance      R2 =' ,R2:5:2,' ohms') ;
   writln('Primary inductance        L1 = ' ,L1:6:3,' henrys') ;
   writeln('Secondary inductance      L2 =' ,L2:6:3,' henrys') ;
   writeln('Mutual inductance         M = ' , M:6:3,' henrys') ;
   writeln('Rotor moment of inertia   J =' ,J:6:3,' kg m2') ;
   writeln('Applied DC voltage        V =' ,V:5:1,' volts') ;
   writln('Initial rotor speed       Wr =' ,w_ri:6:1,' rad/s') ;
   writeln;
```

```
  writeln('Integration parameters: ') ; writeln;
  writeln('Integration step length    =' ,dt:6:3,' s') ;
  writeln('Communication interval     =' ,comint:6:3,' s');
  writeln('Total integration time     =' ,tmax:5:2,' s');
  writeln
end; {of write_parameters}

procedure write_values;
{writes out the variable values}
var i:integer;
begin
  write(t:7:2);
  for i:=1 to nvar do write(x[i]:8.2) ;
  writeln
end; {of write_values}

procedure deriv(t:real; var x,px:array[1..nvar] of real);
{evaluates derivatives required for solution of the induction motor equations}
var
  e_f,e_g,e_d,e_q,i_f,i_g,i_d,i_q,w_r: real;
begin
  i_d:=x[1] ; i_q:=x[2] ; i_f:=x[3] ; i_g:=x[4] ;w_r:=x[5] ;
  e_d := -R2*i_d - w_r* (M*i_g + L2*i_q) ;
  e_q := -R2*i_q + w_r* (M*i_f + L2*i_d) ;
  e_f := -R1*i_f + v_f ;
  e_g := -R1*i_g;
  px[1] := A*e_d + C*e_f;
  px[2] := A*e_q + C*e_g;
  px[3] := C*e_d + B*e_f;
  px[4] := C*e_q + B*e_g;
  px[5] := MJ* (i_d*i_g - i_f*i_q) ;
end; {of deriv}

procedure trapint(dt:real;n:integer;var t:real ;
                  var x,px,w,pw:array[1..nvar] of real) ;
{advances the integration of a set of equations
by one step, using the trapezoidal algorithm;
parameters:
n : number of dependent variables
dt: time step length
t : independent variable (time)
x : array of dependent variable values
px: array of derivative values
w : working array of dependent variable values
pw: working array of derivative values}
```

APPENDIX C: DYNAMIC BRAKING OF AN INDUCTION MOTOR   213

```
var
   i:integer ;
begin
   deriv(t,x,pw) ; t:=t+dt;
   for i:=1 to n do w[i] :=x(i) + dt*pw[i] ;
   deriv(t,w,px) ;
   for i:=1 to n do x[i] := x[i] + dt*(pw[i] +px[i] )/2
end; {of trapint}

begin
   get_parameters;
   writeln('Induction motor dynamic braking ') ;
   writeln; writeln:
   write_parameters;
   writeln(' time' : 7,'i_d' : 7 ,'i_q' : 8,'i_f' : 8,'i_g' : 8, 'w_r' : 8);
   writeln;
   for i:=1 to 4 do x[i] :=0.0; x[5] :=w_ri; t:=0.0;
   write_values;
   repeat
      for i:=1 to steps do trapint(dt,nvar,t,x,px,w,pw) ;
      write_values
   until (t>=tmax) ;
end.
```

Sample output

Induction motor dynamic braking

Machine parameters:

Primary resistance	R1 = 1.80 ohms
Secondary resistance	R2 = 1.81 ohms
Primary inductance	L1 = 0.249 henrys
Secondary inductance	L2 = 0.249 henrys
Mutual inductance	M = 0.240 henrys
Rotor moment of inertia	J = 0.011 kg m2
Applied DC voltage	V = 20.0 volts
Initial rotor speed	Wr = 200.0 rad/s

Integration parameters:

Integration step length	= 0.005 s
Communication interval	= 0.050 s
Total integration time	= 2.00 s

time	i_d	i_q	i_f	i_g	w_r
0.00	0.00	0.00	0.00	0.00	200.00
0.05	−8.87	0.32	9.21	0.01	197.52
0.10	−8.73	0.34	9.07	−0.01	194.31
0.15	−8.73	0.35	9.07	−0.02	191.05
0.20	−8.73	0.35	9.07	−0.02	187.73
0.25	−8.73	0.36	9.07	−0.02	184.36
0.30	−8.73	0.37	9.07	−0.02	180.92
0.35	−8.73	0.38	9.07	−0.02	177.42
0.40	−8.73	0.38	9.07	−0.02	173.85
0.45	−8.72	0.39	9.07	−0.02	170.20
0.50	−8.72	0.40	9.07	−0.02	166.47
0.55	−8.72	0.41	9.07	−0.02	162.66
0.60	−8.72	0.42	9.07	−0.03	158.76
0.65	−8.72	0.44	9.07	−0.03	154.76
0.70	−8.72	0.45	9.07	−0.03	150.66
0.75	−8.72	0.46	9.07	−0.03	146.44
0.80	−8.71	0.48	9.07	−0.04	142.10
0.85	−8.71	0.50	9.07	−0.04	137.62
0.90	−8.71	0.52	9.07	−0.04	132.99
0.95	−8.70	0.54	9.06	−0.05	128.20
1.00	−8.70	0.56	9.06	−0.05	123.23
1.05	−8.69	0.59	9.06	−0.06	118.06
1.10	−8.69	0.63	9.06	−0.07	112.65
1.15	−8.68	0.67	9.06	−0.08	106.98
1.20	−8.67	0.71	9.05	−0.09	101.00
1.25	−8.65	0.77	9.04	−0.11	94.66
1.30	−8.62	0.84	9.03	−0.14	87.90
1.35	−8.59	0.93	9.02	−0.17	80.62
1.40	−8.53	1.05	8.99	−0.22	72.69
1.45	−8.44	1.21	8.93	−0.28	63.94
1.50	−8.26	1.44	8.83	−0.38	54.14
1.55	−7.91	1.75	8.62	−0.51	42.96
1.60	−7.20	2.10	8.16	−0.64	30.15
1.65	−5.81	2.21	7.29	−0.53	16.02
1.70	−3.81	1.42	6.22	0.30	2.93
1.75	−2.40	−0.18	5.97	1.51	−4.10
1.80	−2.04	−0.79	6.63	1.51	−3.69
1.85	−1.81	−0.39	7.18	0.72	−1.42
1.90	−1.52	−0.11	7.53	0.28	−0.46
1.95	−1.27	−0.04	7.79	0.14	−0.19
2.00	−1.06	−0.02	8.01	0.08	−0.10
2.05	−0.88	−0.01	8.19	0.05	−0.06

Bibliography

The following lists are not exhaustive, and are intended as suggestions only.

Background reading

P. Hammond, *Electromagnetism for Engineers*, 2nd ed. (Oxford: Pergamon Press, 1978).

G. Stephenson, *Mathematical Methods for Science Students*, 2nd ed. (London: Longman, 1973).

G. Williams, *An Introduction to Electrical Circuit Theory* (London: Macmillan, 1973).

Further reading

A. E. Fitzgerald, C. Kingsley Jr. and S. D. Umans, *Electric Machinery*, 4th ed. (New York: McGraw-Hill, 1983).

P. Hammond, *Applied Electromagnetism* (Oxford: Pergamon Press, 1971).

N. N. Hancock, *Matrix Analysis of Electrical Machinery*, 2nd ed. (Oxford: Pergamon Press, 1974).

J. Hindmarsh, *Worked Examples on Electrical Machines and Drives* (Oxford: Pergamon Press, 1980).

J. Hindmarsh, *Electrical Machines and their Applications*, 4th ed. (Oxford: Pergamon Press, 1984).

C. W. Lander, *Power Electronics* (New York: McGraw-Hill, 1981).

P. Lorrain and D. R. Corson, *Electromagnetism* (San Francisco: Freeman, 1979).

J. E. Parton, S. J. T. Owen and M. S. Raven, *Applied Electromagnetics*, 2nd ed. (London: Macmillan, 1985).

G. R. Slemon and A. Straughen, *Electric Machines* (Reading, Mass.: Addison-Wesley, 1980).

B. Williams, *Power Electronics* (London: Macmillan, 1986).

Answers to Problems

Chapter 1

1.1. 2980 N.
1.3. The machine will not work.
1.4. Use Ampère's circuital law and the reciprocal property of mutual inductance.
1.5. 0.5 H; 7.76 A; 0.193 H; 1790 N.
1.6. $-\frac{1}{4}L_2 I_m^2 \sin 2\phi$.
1.7. $A = 1.14$; $B = 0.0052$; hysteresis loss 57 W; eddy current loss 13 W.

Chapter 2

2.1. 90.9 per cent; 1.08 Ω.
2.2. $T = \dfrac{KV^2}{(R + K\omega_r)^2}$
2.4. (a) 20 A; (b) 90 rad/s; (c) 20 A; (d) 10 A; (e) 95 rad/s; $\dfrac{d\omega}{dt} + 40\omega = 3800$.

Chapter 3

3.2. 1:3.
3.4. $v_1 i_1 = v_2 i_2$; $Z_1 = k^2/Z_2$; inductance of value $k^2 C$.
3.5. $R_c = 739$ Ω; $X_m = 193$ Ω; $R_e = 0.127$ Ω; $X_e = 0.139$ Ω; $n = 3.04$;
(a) 659.9 V; (b) 97.41 per cent; (c) 1.38 per cent.
3.6. (a) Excessive magnetising current will burn out the transformer.
(b) Very low magnetising current, eddy-current loss unchanged, hysteresis loss 0.233 times normal value.

Chapter 4

4.1. $F(0)$ is the displacement of the θ axis to give equal positive and negative areas.

4.4. $T = \frac{1}{2}K_m^2 \pi r^3 l A_1 \sin\{2(\omega - \omega_r)t + 2\alpha\}$.

Chapter 5

5.1. 36 kW; 500 V; 58.3 A; 0.857; 45 kW.

5.3. $f = \dfrac{1}{2\pi} \sqrt{\left(\dfrac{pT_0}{J \tan \delta_0}\right)}$.

5.4. By Lenz's law, the rotor oscillations will be damped.

Chapter 6

6.2. $T = \dfrac{mp}{\omega} \cdot X_m I^2 \cdot \dfrac{1}{sX_m/R_2' + R_2'/sX_m}$

6.5. Effective rotor resistance R_2'/s in the first case, $R_2'/(2-s)$ in the second.

6.6. Two equivalent circuits in series, with element values halved.

Chapter 7

7.3.

$$\begin{bmatrix} 0 \\ 0 \\ v_d \\ v_q \\ v_f \end{bmatrix} = \begin{bmatrix} R_{kd} + L_{kd}p & 0 & M_{akd}p & 0 & M_{fkd}p \\ 0 & R_{kq} + L_{kq}p & 0 & M_{akq}p & 0 \\ M_{akd}p & M_{akq}\omega_r & R_a + L_d p & L_q \omega_r & M_{af}p \\ -M_{akd}\omega_r & M_{akq}p & -L_d \omega_r & R_a + L_q p & -M_{af}\omega_r \\ M_{fkd}p & 0 & M_{af}p & 0 & R_f + L_f p \end{bmatrix} \begin{bmatrix} i_{kd} \\ i_{kq} \\ i_d \\ i_q \\ i_f \end{bmatrix}$$

Index

AC generators 80, 82 (*see also* Synchronous generators)
Accelerometer, force-balance 7
Alignment torque 18, 112
Alternator (*see* Synchronous generators)
Ampère's circuital law 30
Analogies, electric and magnetic circuit 31
Armature reaction 57
Armature windings 50, 133

Back EMF 64
Breakdown torque 170
Brush contact loss 61
Brushes 45, 52, 55, 61, 184
Brushless DC motor 77
Brushless excitation 131

Cage rotor 157
Capacitor, synchronous 140
Capacitor motors 177
Capacitor-start motors 180
Chopper control 70, 151
Commutation 55, 57
Commutator 45, 50
Commutator transformation 190
Compensating winding 57
Compensator, synchronous 140
Compound motor 69
Controlled rectifier 70
Core loss 61, 95, 135, 173
Coupled circuits 11, 14
Cross-field machines 1, 186, 193
Current density 5
Current density, linear 105

Damper windings 138
DC generators 61
   permanent-magnet 62
   separately excited 62
   shunt 62

tachogenerator 62, 75
transient performance 73
DC machines 45 (*see also* DC generators *and* DC motors)
   armature reaction 57
   armature windings 50
   brush contact loss 61
   commutation 57
   commutator 45, 50
   compensating winding 57
   constants 54, 56
   core loss 61
   cross-field 1, 186, 193
   disc type 75
   dynamic equations 72
   efficiency 61
   elementary 45
   energy conversion 58
   field winding 55
   fundamental principles 46
   general equations 54, 186
   generated voltage 46, 49, 54
   heteropolar 46
   homopolar 7, 45
   interpoles 55
   linear approximation 56
   losses 58, 60
   magnetic forces 57
   magnetisation curve 56
   Maxwell stress 57
   moving-coil 75
   multi-pole 54
   open-circuit characteristic 56
   permanent-magnet 48, 62, 66
   printed-armature 75
   reactance voltage 55
   rotational loss 61
   sign convention 59
   slotted armature 57
   special machines 75
   time constants 73, 74
   torque 50, 54

INDEX

DC motors 64
   back EMF 64
   brushless 77
   compound 69
   disc type 75
   dynamic analysis 73
   electronic control 70
   ideal characteristics 65
   linear 76
   losses 71
   moving-coil 75
   permanent-magnet 66
   printed-armature 75
   separately excited 65
   series 68
   shunt 66
   speed control 66, 67, 70
   starting 71, 73
   torque/speed characteristics 67-9
   transient performance 73
Delta connection 84
Depth of penetration 28
Direct axis 144, 186
Direct-axis reactance 144, 195
Direct-on-line (DOL) starting 71, 171
Disc motor 75
Distributed windings 104 (*see also*
   Sinusoidally distributed windings)
Dynamic analysis of DC machines 72

Eddy current loss 26
Electrical angle 123
Electromagnet 17
Electromagnetic forces 14
   energy methods 19
   Maxwell stress 15
   on a conductor 4
   on iron parts 16, 20
Electromagnetic induction 9
   calculation 13
   Faraday's law 12
   inductance 13
   moving conductor 2
Electromechanical energy conversion 6
Electronic control
   DC motors 1, 73
   induction motors 1, 175
   stepper motors 149
   switched reluctance motors 154
   synchronous motors 131
Energy, magnetic field 19
Energy and inductance 21

Energy conversion 6, 58
Energy methods 19
Equivalent circuit
   induction machine 162, 166
   synchronous machine 134, 135
   transformer 92, 96
Excitation voltage 134
Excitation winding 55, 131
Exciter 131

Faraday 1
Faraday disc machine 7
Faraday's law 12
Ferromagnetic materials 23
Field winding 55
Flux, definition 10
Flux cutting rule 3
Flux linkage 10
Force, electromagnetic (*see* Electro-
   magnetic forces)
Fractional slip 160
Friction and windage loss 61
Fringing 35

Generalised machine theory
   AC machine equations 188
   applications 193
   commutator model 184
   commutator transformation 190
   DC machine equations 186
   direct axis 186
   impedance matrix 188
   primitive machine 184
   quadrature axis 186
   salient-pole synchronous machine 194
   slipring model 186
   torque equation 192
   variable inductance coefficients 188
   voltage equation 187, 189, 191
Generated voltage
   AC machines 45, 81
   DC machines 46, 49, 54
Generator, elementary 7, 80, 82
Generators (*see* AC generators *and*
   DC generators)
Gyrator 101

Harmonics 128
Homopolar machines 7, 45
Hysteresis 23
Hysteresis loss 24

Ideal transformer 87
Impedance matrix 188
Impedance transformation 89
Induced electric field 3
Induced EMF (*see* Electromagnetic induction)
Inductance 11
 energy 21
 induced EMF 12, 13
 leakage 92
 mutual 11
 self 11
Induction, electromagnetic (*see* Electromagnetic induction)
Induction generators 157, 169
Induction machines 157 (*see also* Induction generators *and* Induction motors)
 braking 169, 182, 198, 210
 construction 157
 core loss 173
 dynamic braking 198, 210
 electromagnetic action 159
 equivalent circuit 162, 166
 leakage reactance 165
 magnetising current 165
 magnetising reactance 165
 multi-pole 166
 power relationships 167
 rotational loss 173
 rotor efficiency 168
 slip 160
 slip frequency 161
 torque 161
 torque/speed characteristic 168
Induction motors 157
 applications, single-phase 180
 breakdown torque 170
 cage-rotor 157
 capacitor 177
 capacitor-start 180
 characteristics 171
 efficiency 171
 electronic control 175
 Kramer system 174
 plugging 169
 pole-amplitude modulation 175
 pole-change windings 174
 power factor 171
 rotor resistance, effect of 170
 shaded-pole 179
 single-phase 177

 slipring 157, 174
 speed control 173
 split-phase 179
 starting 171
 torque/speed characteristic 170
 variable frequency 175
 wound rotor 157
Instrument transformers 96
Interpoles 55
Inverter 1, 131, 176

Kramer system 174
Kron 184

Laminations 27
Leakage 35
Leakage flux 91
Leakage inductance 92
Leakage reactance 94, 135, 165
Lenz's law 12
Linear current density 105
Linear machines
 DC 76
 induction 180
 reluctance 146
 synchronous 145
Linearity, magnetic 11, 35
Load angle 137
Lorentz equation 2
Losses (*see* Core loss, Eddy current loss, Hysteresis loss *and* Rotational loss)
Loudspeaker, moving-coil 7

Magnetic circuit 28
 analogies 31
 calculations 32
 diagram 32
Magnetic field energy 19
Magnetic flux 10
Magnetic fringing 35
Magnetic leakage 35
Magnetic linearity 11, 35
Magnetic materials 23
Magnetic potential difference 30
Magnetic saturation 24, 34, 56, 63, 65, 185
Magnetisation curve 23
Magnetising current 44, 135, 165
Magnetising reactance 95, 135, 165
Magnetomotive force 31
Maxwell stress 15, 57, 111, 203

Meter, moving-coil 7
Motional induction 2
Motional induction formula 3
Motor, elementary 7
Moving-coil
  DC motor 75
  loudspeaker 7
  meter 7
Multi-pole machines 54, 122, 136, 166
Mutual inductance 11

Negative phase sequence 86

Oersted 1
Open-circuit characteristic 56
Open-circuit test 98

Penetration depth 28
Permanent magnets 23, 38
  Alnico 38, 40
  coercivity 39
  demagnetisation characteristic 39
  ferrite 38, 40
  neodymium–iron–boron 38, 39
  rare-earth 38
  recoil 40
  remanence 39
  samarium–cobalt 38, 40
Permeability 23
Permeance 38
Phase sequence 86
Plugging 169
Pole-amplitude modulation 175
Pole-change motors 174
Polyphase systems 82
Pony motor 143
Positive phase sequence 86
Potential difference, magnetic 30
Power electronics (*see* Electronic control)
Power-factor correction 140
Primitive machine 184
Pull-out torque 138, 153

Quadrature axis 144, 186
Quadrature-axis reactance 144, 195

Reactance voltage 55
Rectifier, controlled 70
Regulation 99
Reluctance 31
Reluctance motors 145

Reluctance torque 144
Reversal of rotating field 117
Rotary amplifiers 1
Rotating magnetic field 114
  definition 114
  induced voltage 118
  multi-pole 122
  reversal of direction 117
  space and time phasors 120
  three-phase winding 116
  two-phase winding 115
  voltage–current relationship 119
Rotational loss 61, 173

Salient poles 103, 143
Saturation 24, 34, 56, 63, 65, 185
Schrage motor 1
Scott transformer connection 100
Self-inductance 11
Separately excited generators 62
Separately excited motors 65
Separation principle 75
Series motors 68
Shaded-pole motors 179
Short-circuit test 98
Shunt generators 62
Shunt motors 66
Sign convention 12, 59
Silicon steel 23
Sinusoidally distributed fields 104, 106 (*see also* Rotating magnetic field)
  combination 108
  space phasor 110
  vector representation 110
Sinusoidally distributed windings 107
  current density 106, 107
  rotating magnetic field 114
  torque 111
Skin depth 28
Skin effect 28
Slip 160
Slip frequency 161
Slipring induction motors 157, 174
Sliprings 45, 131, 157, 184
Slotted armature 57
Soft start 71, 171
Space phasors
  definition 110, 111
  relationship with time phasors 120
Speed control (*see* DC motors *and* induction motors)

INDEX

Split-phase motors 179
Squirrel-cage rotor (*see* Cage rotor)
Star connection 84
Star–delta starting 171
Star–delta transformation 85
Starting
   DC motors 71, 73
   induction 171
   synchronous motors 143
Steinmetz law 24
Stepper motors 131, 146
   bipolar drive 151
   chopper drive 151
   electronic control 149
   hybrid 149
   multi-stack 147
   multi-step operation 152
   permanent-magnet 147
   pull-out torque 153
   resonance 152
   single-stack 147
   slewing 153
   start/stop rate 153
   static torque 151
   switched reluctance 154
   unipolar drive 149
   variable-reluctance motor 147
   velocity profile 154
Stress, Maxwell 15, 57, 111, 203
Stress vector 16
Superconducting machines 1
   DC 9
   synchronous 131
Synchronous capacitor 140
Synchronous compensator 140
Synchronous generators 131, 139
Synchronous machines 131
   armature winding 133
   brushless excitation 131
   characteristics 140
   core loss 135
   damper windings 138
   direct-axis reactance 144, 195
   equivalent circuit 134, 135
   excitation voltage 134
   excitation winding 131
   leakage reactance 135
   load angle 137
   magnetising current 135
   magnetising reactance 135
   multi-pole 136
   phasor diagrams 131–4
   pull-out torque 138
   quadrature-axis reactance 144, 195
   reluctance torque 144
   salient-pole 103, 143
   superconducting 131
   synchronous reactance 135
   synchronous speed 136
   synchronous torque 137
   torque/load-angle characteristic 137, 143, 144
   two-axis equations 195
Synchronous motors (*see also* Synchronous machines)
   electronic control 131
   linear reluctance 146
   linear synchronous 145
   power-factor correction 140
   pull-out torque 138
   reluctance motors 145
   starting 143
   V-curves 140
Synchronous reactance 135
Synchronous speed 136

Tachogenerators 62, 75
Three-phase systems 83
Three-phase to two-phase transformation 206
Thyristor control 1, 70
Time constants, DC machine 73, 74
Torque
   alignment 18, 112
   breakdown 170
   calculation 111
   DC machine 50, 54
   induction machine 161
   pull-out 138, 153
   reluctance 144
   synchronous 137
Transformation
   commutator 190
   impedance 89
   ratio 88
   star–delta 85
   three-phase to two-phase 206
Transformer induction 9
Transformers 86
   auto-transformer 100, 171
   construction 87
   core loss 95
   determination of parameters 97
   efficiency 99

INDEX

equivalent circuit 92, 96
ideal 87
impedance transformation 89
instrument transformers 96
leakage inductance 90
leakage reactance 94
magnetising current 94
magnetising reactance 95
open-circuit test 98
phasor diagram 95
primary 87
ratio of transformation 88
regulation 99
secondary 88
short-circuit test 98
tests 97
Transient performance
   DC generators 73
   DC motors 73

induction motors 198, 210
Two-axis equations 195

Universal motors 68

Variable-frequency induction motor 175
V-curves 140
Voltage (see Electromagnetic induction and Generated voltage)

Ward–Leonard system 62
Windage and friction loss 61
Windings (see also Sinusoidally distributed windings)
   AC machine 124
   DC machine 50
Wound-rotor induction motors 157, 174